BRAINWORMS

AND OTHER SURPRISES AT THE MARGIN OF THE IMPOSSIBLE

BY ALBERT DRIEDGER

FriesenPress

One Printers Way
Altona, MB R0G 0B0
Canada

www.friesenpress.com

Copyright © 2024 by Albert Driedger, MD, PhD, FRCPC, FACP, FACPE
First Edition — 2024

Editor — Kaitlyn Sutey

ISBN
978-1-03-831944-9 (Hardcover)
978-1-03-831943-2 (Paperback)
978-1-03-831945-6 (eBook)

1. SCIENCE, ESSAYS

Distributed to the trade by The Ingram Book Company

Dedication

This work is dedicated to the memory of Lilianne, whose gentle criticisms and consistent good humour helped shape parts of these reflections, and whose graciousness and constant love sustained me through more than sixty years of life together.

"If you aren't in over your head, how do you know how tall you are?"

— T.S. Eliot

TABLE OF CONTENTS

BRAINWORMS

WHO AM I?

I am a retired eighty-seven-year-old physician-scientist, now Emeritus Professor at Western University. Over the past twenty years, many family members, and even some friends, have urged me to write a biography, because it seems to them that I have led an interesting and eclectic life. I tried to do that, but the temporal linear version was so damn boring because I knew how it would end. Good writing should surprise the writer as much as the reader, especially when it touches on truth. I've always lived more in my mind than in the world, and I began to write these little editorials to myself, which I named "Brainworms." When they eventually emerged toward a widening readership, I summoned the courage to let a little sunshine onto their content. From time to time, there will be glimpses of me in these pages, but not in a linear fashion—more like the tumbles of a kaleidoscope. Occasionally, I get a bit of vigorous feedback, and I love the opportunity to debate without rancour; a skill that our society is losing.

My beginning was not auspicious; I was born on a mid-nineteen-thirties dust patch, euphemistically called a farm, in Saskatchewan. Everything on the property was overlaid by the patina of poverty. Who would have thought that advancement above the bare necessities of survival was even possible for me!

The central professional piece of my life became nuclear medicine, after I had already completed training in internal medicine and a second doctorate in biochemistry. In the middle of my career, there was a five-year period with an emphasis on health system management. For thirteen years, I consulted with government agencies and industry groups on issues related primarily to health and safety in the nuclear energy industry. For another twelve years, I episodically taught a course concerning physicians' roles in hospital management in the Middle East and in Sudan. In my retirement at age seventy, I became involved in environmental matters and assisted at Thames Talbot Land Trust, an NGO, in the management of a local challenged forest environment Until age eighty-three, I could be found in the Five Points Forest near my home, swinging a chain saw at invasive trees. I recently gave up my chain saw and now live in a condo in Huntsville, Ontario, with my cat. I write a bit when inspiration overcomes my reluctance to be found out.

WHAT IS A BRAINWORM?

Is it the same as a "brain worm?" We were all grossed out to our shuddering bone marrows in 2023 by the story of the Australian woman who presented with bizarre symptoms and was found to have a fifteen-centimetre-long roundworm—the kind that belongs in a python's belly—lodged in her brain. How did that one get mislaid so badly? Is that the only worm that might, however improbably, infest the human brain? Be reassured: there's more.

Also intriguing, but not as immediately disgusting, is the *Toxoplasma gondii*, a small gut parasite of many cats. There is not much to see here. It lives comfortably with its host and lays its eggs into the cat's feces. In the wild, rodents are known to pick up the eggs from contact with cat feces, and they become secondary hosts to the larvae. The surprise is that the infestation causes them to lose their fear of cats. The result is that a random cat encountering the infested mouse has an easy meal, and the next generation of *T. gondii* is homed, thereby continuing the propagation of its kind into the world. This observation is not limited only to our house cats; there is a recent report of hyenas infected by *T. gondii* who similarly lost their fear of lions. One might suppose that the encounter did not go well for the hyena either.

Humans around the world have close associations with house cats, and many become infected with the same *T. gondii*, most often without symptoms, although blood tests will demonstrate the immunological footprint of a previous encounter with the parasite. In this process we also become secondary hosts. Many questions arise. For example, is the parasite "waiting" for us also to be eaten by an improbable cat, so it can get on with its life? Does it also alter our experience of fear? And what are the mechanisms of its action, if any, on our brain? It was hypothesized that there might be increased titres of antibodies against *T. gondii* in people who were injured while engaging in risky behavior. The results were plausible, but the number of people tested was small. I don't know if anyone has undertaken brain imaging of cat lovers to look for the larval cysts. I'm still cautious about accepting the inference that playing with cats in childhood might be a defense against a charge of reckless behaviour in later life. Obviously, the cuteness factor of small children cuddling kittens easily overcomes any remote parasitological considerations.

In this tentative exploration of *T. gondii*'s effect on the brains of secondary hosts, humans are presumably being set up for consumption by a cat. Once, while on a hike through the savannah in Botswana, I asked my unarmed guide about the proper code of conduct if we encountered a lion. His reply invoked what he called the lion's creed, "Food runs." Possibly fearlessness, even if only supported by a parasitic crutch, could be beneficial in this setting. We didn't meet any lions that day.

But none of the above are in any way related to the ramblings I choose to call "brainworms." What I choose to designate with the term pertains to the rumblings and grumblings of an unsatisfied mind confronted by the infinite margins of the unknown. As a child who read everything that I could get my hands on, I worried everything would be discovered before I had a chance to make my own contribution. I needn't have worried; eventually I had opportunities enough. My voyages are all done now, but my curiosity continues to drive me to trawl for novelty, although age limits me to fishing from the shore. The invocation of brainworms is the way I scratch my curiosity, and they daily stoke my awe of the vast and wonderful universe we, however briefly, inhabit. I hope you will enjoy the book. If you do, please pass it on.

HALF AN EYE

One Sunday morning forty years ago, I was in my kitchen drinking coffee and reading a book about evolutionary biology when my youngest preschool-aged daughter, came into the house with a rock she had split behind the barn, and in which she recognized a fossil. It was, indeed, a well-formed Trilobite from about 300 million years ago, and it measured about three inches across. Its most amazing feature to me was the hexagonal lenses of its eyes made of an inorganic calcium phosphate crystal. The amazing coincidence was that I, at that moment, was reading about the evolution of vision, and my book was open at a page with a photo of a Trilobite fossil that displayed the very point. One can, even today, measure the refractive indices of these lenses and determine the limits of visual resolution these creatures had.

Evolution developed many mechanisms for light detection throughout biological history, of which the crystalline lens of the Trilobites was one. Some of the others were the melanotic eye spot of single-celled amoebae; the compound eyes of insects; those of hunting birds, like kestrels, that are sensitive to ultraviolet rays for better prey detection; and many unique others.

We take the property of stereoscopic vision for granted in our own experience of sight, but mammals with eyes arranged more laterally (eg, horses) do not have the necessary overlapping visual fields. The development of these individualized eyes and brains required millions of years.

Some aggressive creationists use rhetorical questions to confront and trip up evolutionary biologists, one of which is, "What use is half an eye?" The mistake in that line of thought is that evolution has never been about development toward a predefined goal, such as an eye, but to use available resources to improve the odds of survival in the existing environment with just one additional random variation from time to time. It is clear now that from the earliest times, life experimented with the challenge of environmental awareness, of which light detection and image management is one important factor. Some fifty to seventy different designs for eyes appear in the fossil record and in the inventory of extant biota, some of which worked in one niche or another and then were abandoned as better options arose in adaptation to changing environments. The trilobite eye never made it into modern biology, as its image resolution was poor, and survival required something better. Today, insects, squids, single-celled amoebae, birds, mammals, and even bacteria all have varied, even fundamentally different, light and image management systems.

Craig Ventor is an imaginative molecular biologist with many fundamental insights to his credit. A while ago, he outfitted an ocean-going yacht for marine collections and went on a world cruise to seek out novel forms of marine microscopic life. Specifically, he studied the regions known as the Sargasso Seas, which were then believed to be devoid of food sources, and thus incapable of supporting life. His search tool was DNA sequencing of serial sea water samples, and he identified some tens of thousands of species living and flourishing in these areas, with variation from place to place. Even in the Sargasso Seas, those so-called, ocean deserts, there were a hundred thousand bacteria and a million viruses in every millilitre of water. It turned out that the bacteria were using a protein, rhodopsin, to scavenge sunlight for their energy needs. Importantly, in vertebrate eyes like ours, this same protein traps photons and generates the electrical impulses that transmit visual signals to the brain. In bacteria without eyes, it served in a different biological niche.

Back to the creationist's taunt: "What good is half an eye? It all depends on the context. To the ocean dwelling bacteria in the Sargasso Seas, it is a full-fledged energy harvester. I don't know how the Trilobites developed their images, or whether rhodopsin was a functioning element in their eyes; their history would be interesting if we could only read all of it. We could quite reasonably speculate that mammalian eyes are a consequence of the microbial origin of rhodopsin. Half an eye isn't everything, but it bests willful blindness.

NO END OF WONDERING

A recent book that will fascinate nature lovers and mystify many of the rest of us is *Entangled Life* by Merlin Sheldrake. The story begins with the excitement of truffle hunting with dogs in the Italian Piedmont region. There is much more about mushroom picking/eating, as well as the psychotropic effects of some species on offer. There is so much more in the book about the life that thrives in the earth under our feet, and that is where I will take you next.

In the laboratory, fungi exhibit some complex behaviours that suggest they might be acting in purposeful ways. You can see this for yourself with an easy experiment. Set up a moist, soil-topped table, with the soil shaped into a map, and with pieces of slightly rotted food-source wood, sized proportional to population. Those pieces of wood should be organized to mark the locations of major cities. When organized in this way, certain species of fungi will quickly grow their hyphae (i.e., their roots) into patterns and sizes that connect the "cities" with routes (roots!), so as to minimize transportation costs (i.e., cheapest possible food) between centres. In this activity, they communicate across metres of space with electrical signals, which resemble nerve cell functions in human brains and convey information about where food is available and needed at any given moment. Transported from the laboratory to the forest, the interconnected underground networks span dozens to hundreds of metres, and they have dozens—even hundreds—of trees in communicating networks. There may be hundreds of somewhat independent networks manipulated by different fungal species.

When the existence of this interconnectedness was first mooted by Suzanne Simard, the editors of the journal *Nature* dubbed it "The Wood-Wide Web." It isn't being suggested that there is any actual "thinking" going on in the ground under our feet, but it would make a neat science-fiction scenario.

We are becoming ever more aware of the essential role that fungi play in their support of trees and the wider ecology of forests. They grow out hundreds of metres-long fibers, aka hyphae, that grow into the roots of many species of trees, where they assume the role of supply chains operating the arboreal ecosystem. For example, in the early spring in British Columbia, when fir trees have access to sunlight, they photosynthesize sugar from atmospheric carbon dioxide, which they share with the still-naked nearby birch. In midsummer, when the broad leaves of the birch shade the fir, the trade is reversed. And it reverses again in the fall, after the birch leaves have fallen and the weather still permits some degree of sugar synthesis to continue in the then again-donor firs.

From other sources, I know that all is not sweet and "light" down there in the dark under our feet. There are also ongoing chemical wars whose sophistication makes what we humans have, so far, done to each other pale in comparison. For example, thirty years ago, I planted a few walnuts in the Five Points Forest near Ingersoll, Ontario. At first, they found themselves in an intense competition with neighbouring mixed growths, but over time, the competitor species died off, as they were progressively poisoned by the juglone excretions of walnut seedlings. There are many other examples of chemical warfare among plants, and in some cases, their chemical vectors, such as the thioglycosides secreted by garlic mustard, are also toxic to fungi.

When I visited Africa, I was dismayed by the messy eating habits of elephants; one would rip a branch out of an acacia tree, only to abandon it partially eaten and then move along the forest to repeat. Then, my guide informed me that the acacia began to manufacture bitter alkaloids at the moment when the branch was being torn off, and that subsequent bites from that tree would be inedible to the elephant for some time. Moreover, nearby trees, whether triggered by odour detection or by electrical signals via interconnecting roots, were also stimulated to produce those protective alkaloids, thus protecting the entire grove. The elephants did move a distance before they found another unsuspecting, sweet acacia. It appeared that the trees were cooperating to repel a raid. Question: were they acting as a cooperating militia of individual trees, or was this behaviour mandated by the interconnected social unit of trees, i.e., the forest?

To return to Sheldrake's book: the author also creates space in which to wonder about the nature of individuality and the spectrum of possibilities for cooperation in a natural system that we think of as driven by Darwinian evolution. The difference between cooperation and competition for survival is a difficult conceptual balance for some biologists. Some observers might conceive of the forest as the collective entity, rather than as individual trees. There clearly is life at the level of the individual plant supported by Mendelian genes, mutations, and a metabolism, but perhaps there is also existence at a supra-Mendelian level—that of the cooperative unit, also known as the forest. Translated to a political level, this might reduce to an argument between individualistic libertarians and welfare-minded socialists, both struggling toward a healthy woodland. It seems necessary to entertain both models, as though they were in a symbiotic union, while reflecting on the distinction between what goes on above and below the ground.

Sheldrake ends his excellent book with a survey of the benefits we can hope to accrue from a better understanding of fungi. Many products that will be totally fungally sourced and totally compostable are being developed for the marketplace; these will range from vegan bacon to leather goods, to a wide range of construction materials, and much more. Here is a website that describes some current business opportunities: https://ecovative.com/.

In any event, Sheldrake's book concludes with one hundred pages of tightly written notes and references, and I leave all of it to your wonderment.

THE SCREAM

A little while ago, I walked a trail that featured a series of hanging footbridges over a dozen mountain valleys in Costa Rica. The forest canopy was about 50 metres below (and I had to wonder: how much further below that to the forest floor?). While looking down, I heard a sound and looked back to see who was coming up behind me. It was the English woman who had earlier proclaimed her fear of heights. The expression on her anguished face connected me directly with Edvard Munch's painting, *The Scream*: eyes wide, eyebrows raised, and mouth agape.

Fear is our most primitive reaction to dire situations. In our brains, fear is generated from two small regions known as the amygdalae (the Latin name for almond). The amygdalae are part of our paleocortex (or "old brain"), which originally served to allow our reptilian ancestors to make unthinking, quick responses to unexpected changes two hundred million years ago and still performs similar functions for us today.

Facial expressions help us read how our fellow humans feel, be it happy, sad, angry, attracted, disgusted, or fearful. All are typically interpretable from a reading of our faces. Normal subjects who are shown pictures of faces expressing these emotions while their brains are being imaged in a functional MRI machine will reliably be shown to have activated their right amygdala. There is only one exception: psychopaths do not recognize fear in others, nor in themselves, and their amygdalae are not metabolically responsive to these stimuli. In a way, it's not surprising that people who are prone to act out their immediate impulses without regard for consequences might also not understand fear. Fear has its own nervous pathways, and it is distinct from other emotions.

It was only recently discovered that fear is uniquely expressed on the human face via unique rapid conducting nervous pathways; this happens so quickly that an observer will recognize the expression on a subject's face before the subject themself knows that fear is called for. Imagine the cartoon scenario of the caveman who spots a lion approaching over his friend's shoulder. Where is the evolutionary survival advantage in the observers' fear-filled face, unless it benefits the friend at risk to take notice of the expression? The schematic drawing of the fear fiber network is as complex as a microchip circuit board. It doesn't matter that you, like me, don't understand it; the important concept is that at an operational level, fear is an electrical circuit.

In contrast to psychopaths, some other people are known for their acts of supreme generosity—e.g. donating an organ to a stranger and without any thought of a benefit—and they characteristically have a hyperactive right amygdala demonstrated on functional MRI images. A heightened fear response also correlates with empathy, and this seems not to be a coincidence. In a crisis, the most frightened person on the scene may also become the hero of the day. Thus, at a socio-psychological level, one must know fear in order to be an empathic person.

It is interesting that a work of art from 1893 would resonate so well with current science.

OF EMPATHY AND ETHICS

I'm reflecting on a debate I heard between Richard Dawkins and John Lennox. Both are, or were, professors at Oxford University; Dawkins is an evolutionary biologist and vigorous atheist, while Lennox is a mathematician and evangelical Christian in the tradition of C.S. Lewis. It was all done in a very proper British way, with Lennox mostly in retreat, throwing out diversionary theological bombshells as he bounced from one topic to another, and Dawkins attempting to hold him to the point of the discussion. They did that sort of dance until the topic of morality came up, when they both ground to a halt in disagreement, but without a solution over the source of the human sense of right and wrong. Lennox opted for a Newtonian, deistic intervention that had God reaching into humanity to install a divinely engineered morality module, which would accept Scripturally encoded instruction and Dawkins professed a mixture of puzzlement and confidence that evolution would get it right. I wanted so much to leap into the debate at that point because I suspect that there can be an evidence-based starting point to this discussion, which they both missed. Because of that, their debate ground to a halt as they groped for final words with which to run down the clock.

The core of my argument deals with the existence of the mirror nervous system (MNS), a collection of brain cells whose properties and interconnectedness makes it possible for us to anticipate the reactions of others. We have learned enough about it to be able to set up testable scenarios of its functions. Initially, studies of the MNS were performed by inserting electrodes into cortical neurons and measuring voltages in the cells of awake animal brains. New observations came quite by accident when an investigator had, for a different purpose, connected a chimpanzee to his recording apparatus and then stopped to think a moment before launching the actual experiment. While pondering his next move, he absentmindedly picked up a peanut from his desk and ate it. The monkey, watching the investigator, projected a signal on the monitoring oscilloscope that had only previously occurred when it enjoyed the peanut itself. The researcher saw the signal and interpreted it as the monkey thinking, "That must be a good peanut that you're eating." In short, the chimp was vicariously enjoying a human experience. This observation first clued scientists into the existence of the MNS. The question then became, "Can we test this in human experience?"

We can take the experiment up a notch from the monkey to a human subject in an fMRI scanner. This is a machine that produces images that show precisely where in the three-dimensional brain a particular signal is processed. With the subject in position, a single stimulus, painful or not, was administered, and the concurrent image of the brain showed where the sensation was processed. Let us suppose that this experimental subject brought her toddler along for her session, and that the scanner operator agreed to supervise the child while Mom was in the scanner, but in the same room. An electrode on the subject's finger administered a single sharp shock, and a brain image was recorded. The electric shock part of the experiment being over, the electrodes were placed on an adjacent table. While the technician was otherwise engaged, the child picked up the electrodes and accidentally also received a shock, which made it to cry out, "Ow, that hurts!" At that instant, Mom's fMRI image took on the same appearance as when she had herself received the shock. In other words, she perceived her child's pain as her own.

Similar results were obtained when other subjects, while in the MR machine, were asked to look at images of hands mangled in industrial accidents. For this study, the subjects had received an anesthetic nerve block in their corresponding hand. Surprisingly, they complained that viewing the photos was painful in the anesthetized hand. They said something like "May I stop looking at this now? It hurts my hand." This demonstrated that the images were actually causing pain-conducting nerves in the brain to fire while nerve conduction from the hand was blocked. Once again, "Someone else's pain hurts me." This is a plausible beginning of an evidence-based ethic, promoting the evolution of empathy. It raises a possibility for progress in understanding of our concepts of right and wrong via the realization that one's security is linked to the welfare of others in the tribe. It might be related to oxytocin (the "love hormone") release. What emerges from this experiment is an expression of the Golden Rule in the context of brain function.

What role does empathy possibly play in evolution? Is cooperation and caring a stumbling block for the theory of survival of the fittest? Not at all. At the lowest possible level, one might worry about a fellow tribesman like one might worry about the status of one's tools. The mutuality of the feeling grows toward friendship and thence becomes loyalty to the pack or tribe. Our literature is replete with heroic tales of people who stood in harm's way at great personal cost to protect other people or a community's values. The survival (or not) of the hero at these crisis points is not the issue; it is about survival of the best genes, which will help ensure survival of the group.

It has become clear to me that we do not need to hope for the insertion of a divinely ordained ethics module *à la* Professor Lennox so that we may know good from evil. We are not lost, but only finding our way, and there is a difference. I would like to hear the esteemed professor's response to that.

"THE QUALITY OF MERCY IS NOT STRAINED"

I took some time this week to acquaint myself with recent developments in the field of artificial intelligence, only to discover that we have moved beyond first-generation AI and on to computers that program themselves by observing and learning. I saw a video of such a robot in an operating room. The robot learns, by observing the human activity, to reach into the operating field and hold the surgeon's thread with just the correct degree of tension to tie the best knots. Underlying these astonishing capabilities is something I only vaguely comprehend: a neural network. But how does one 'teach' a robot its business? This is done by exhibiting very large databases of the best human performance at the designated task; we do not expect the robot to exceed human performance—only to approach it. There is, of course, the positive factor in that the robot will not become fatigued or distracted by external issues.

Some of the impetus to develop robots has come from public security concerns, such as the facial recognition of terrorists, but the same techniques can also be applied to other tasks, such as the analysis of diagnostic medical images. If I were a young physician today, I would look about for something more than diagnostic radiology or microscopic pathology to build my career on because that work will increasingly be done by robots. These technologies should serve to warn physicians away from specializing too narrowly

19

in some of the technological developments in medicine and to stay close to direct patient contact. The era of medical images as quasi video games may be coming to an end as the video game learns to play itself.

Another area where deep learning is revolutionizing the field is that of language interpretation. We remember when the inability of computer translations to master idioms was the butt of bad jokes. It seems that the capacity of computers to learn by hearing rather than by programming has surmounted this difficulty to the extent that spontaneous, idiomatically correct translation is possible. These devices allow you to speak English and hear yourself speaking idiomatically correct Chinese or Greek—without an accent, in your own voice, and with proper pauses for your breathing! I haven't heard whether the robot can translate your lame English-language jokes into culturally appropriate Chinese, though.

It occurred to me as I was learning all this that there are some things that we humans feel obliged to do, but do very badly, whereas robots might excel at the very same task. I am thinking, for instance, of evaluating the probabilities of future human behaviour; eg, calculating the risks inherent in granting parole to prisoners.

Daniel Kahneman won a Nobel Economics Prize for work he did largely through carefully constructed questionnaires and observations of human behaviour. One of those experiments concerned the process of evaluating parole applications from prison in Israel. The process was that three of the country's most senior judges met a few times each year to consider applications; in order to complete the work in hand, they had to decide every three–five minutes. Kahneman gained access to the session under the pretense that he was studying some aspects of the applicants' personalities. Actually, he recorded the time of day at which each application was decided in relation to the timing of the judges' tea and lunch breaks. The judges began the day in a generous way and initially granted most requests but, as they became fatigued, fewer paroles were granted. At midmorning, they stopped for tea and a biscuit, and immediately following that, they again granted 90% of the applications. In the final half hour before lunch, the pass/fail ratio was again reversed. For details, see Kahneman's book, *Thinking Fast and Slow*.

I think we ought to release criminals as soon as we are confident that they will not endanger us. In Canada, it costs six figures to incarcerate one person for a year, and timely release serves all our interests, in part through the efficient use of prison resources. The evidence shows that parole boards may not be effective in their present way of considering requests for parole. Perhaps a computer capable of learning from experience would do better. For one thing, the robot-decider could use databases incorporating experiences on the actual incidence of recidivism among similar antecedent parolees. I think it unreasonable that the quality of mercy should simply flow through a judge's tea strainer.

DOES ANYONE REMEMBER LEWIS THOMAS?

Lewis Thomas became one of my favorite physician models when I was a medical resident and in the years that followed, when he wrote occasional columns in the *New England Journal of Medicine* under the headline of "Notes of a Biology Watcher." These columns featured eclectic musings of catholic taste, including medicine, history, linguistics, questions such as the relative intelligence of individual bees versus that of the hive, and many other speculations. He held my full attention when he wrote about his experience as a hospital intern in 1937, the year in which I was born. He summarized that time as a year in which, despite assiduous attention to duties, he had not managed to alter the natural course of disease in even a single patient, nor had he been expected to do so, because the potential for physicians to treat diseases with the intent of curing the patient was not yet in clear focus. Not content to consider himself ineffective, he opted for a career in pathology instead of a directly patient care-related specialty.

It comes as a shock today to realize that medical intervention with the objective of cure is a relatively recent innovation; some in the not-so-distant past would have said it was also irreverent. The Hôtel-Dieu in Paris, aka "God's Hostel," began to provide shelter for a mélange of homeless, sick, and starving people in the seventh century and has morphed over time into a modern hospital that at times housed over 4,000 sufferers. Although charity and compassion were practiced in abundance in such institutions, it seems not to have occurred to the hospital's founders or its subsequent managers that anyone would have benefitted from attempts to actually treat diseases; indeed, it might be irreverent to impede the manifest will of God. Physicians were present in abundance in these hospitals, but they seemingly were more interested in making correlations of symptoms and signs in life with findings at autopsies. The very notion of cures for any disease would have been seen as arrogant—if it could even be entertained—because suffering was to be endured.

For instance, as late as the mid 19th century, tuberculosis was defined as a "congenital disease of the working class." Where, in that formulation of disease, was there any—even *implicit*—mandate to ease suffering? Empathy is a resource that is quickly exhausted when confronted by futility, especially if it is disguised as "God's Will." When Edward Jenner demonstrated in 1796 that smallpox could be prevented by inoculation with the cowpox, he was roundly abused by the church for tampering with "Divine intent." It seems that the gradual appearance of effective remedies from about the mid-nineteenth century greatly encouraged positive attitudes toward really caring for the sick. No better example can be found to support this point than Florence Nightingale, who threw herself and her small band into cleaning up the desperately filthy British Army hospital in the Crimean War and thereby reduced the death rate among her patients by two thirds in one year. Her inspiration came in part from the emerging understanding of the role of bacteria in wound infections; she realized that cleaning up the putrid messes would benefit the soldiers.

In the late nineteenth to early twentieth century, Paul Ehrlich noted that there were chemical dyes used in the laboratory that stained tissue specimens in diagnostically useful patterns. He speculated that it might also be possible to find other chemicals that would bind to diseased cells with therapeutic effectiveness and also

23

identified syphilis as a major cause of disease at the time. (Indeed, the stages of the disease manifested themselves in every organ, and the early practitioners of what came to be internal medicine were first known at that time as syphilologists.)

In the nineteenth century, art and literature typically depicted the exhausted doctor sitting at the bedside through the night to determine the course of the disease and interpreting clinical signs to the family hovering in the background but without any therapeutic options to offer. As Lewis Thomas noted, therapeutic medicine had a slow start. As we slowly moved beyond potions, notions, nostrums and secret remedies toward mechanistically sound understandings of the interactions between diseases and scientifically derived medications, new roles also evolved for physicians with evermore possibilities for the communication of realistic hope to patients and families based on evermore rigorous understandings and enhanced skills. Never did we, either as physicians or patients, have so much to live for, and I am grateful to Lewis Thomas for the stimulating thoughts that he dropped into my mind.

"BY THEIR FRUITS SHALL YE KNOW THEM"

The essential link between underlying assumptions and outcomes is often overlooked, especially when an ideological framework has been imposed beforehand. Just in case you're wondering, you know you have an ideology underfoot when you are told what the answer must be before you have fully learned how to formulate the question. Nowhere is this better or more tragically illustrated than in the case of Soviet-era biomedical research and their parallel mismanagement of agriculture. The regime's ideologues supposed that their system inherently contained the solutions they needed without reference to evidence. In this interval, the western world discovered antibiotics, cancer treatments, and therapies for hypertension, heart disease, and multiple other maladies, as well as molecularly specific approaches to food production and many agricultural issues. How many biomedical or agro-biological advances came to us from the Soviet Union? The answer is that there were none.

At the beginning of the era, Russian scientists were not separated from the west by any unbridgeable knowledge or technology gaps, and where it mattered to them, as in the arms race, they were quite able to keep up with us. How did it happen that they lagged ever farther behind in the biomedical and agricultural sectors? The answer has several parts, including their leaders' total lack of empathy for the needs of the population and a reciprocal wariness on the part of their professionals to avoid crossing ideological red lines, particularly as doing so became ever more costly in personal terms.

When the dust settled after the Bolshevik Revolution, Stalin sought a philosophical underpinning to justify his tyranny. He needed an atheistic philosophy, but at the same time, he rejected emerging views of Darwinian evolution, since they couldn't be used to justify his program of property collectivization and social repression. One of his advisors, Trofim Lysenko, was a plant scientist who espoused an older view of evolution first enunciated by Jean-Baptiste Lamarck in the early nineteenth century. Lamarck, and consequently Lysenko, thought that life forms transmitted acquired characteristics to their offspring; for instance, one could evolve plough horses toward race horses by training them before breeding. Stalin saw in this view the opportunity to justify oppression as a form of training to progressively breed undesirable traits, such as capitalism, out of human society. Thereafter, he argued that the race would spontaneously support the greater good, and government would atrophy in accord with Karl Marx's prediction.

In Stalin's Soviet Union, any scientist who expressed support for Darwinian evolution was guaranteed to be branded as subversive and deserving of a one-way ticket to the Gulag, or even execution. Predictably, biologists learned not to ask questions or to follow up on experiments that might yield dangerous insights. The Darwinian approach to biology was no longer taught in their schools or universities. For Soviet biomedical scientists, the absence of a unifying hypothesis that linked animal and human life broke the link that, for western scientists, justified animal experimentation to predict human responses. Without that critical unifying theory, animal research would have been irrelevant to human medicine.. Thus, the Soviet Union made essentially no contribution to international standards of health care.

Similarly, Stalin's rejection of Darwinian evolution also prevented the application of genetic science to agriculture. Lysenko's misinformed biology led to wholesale crop failures, and evidence of his incompetence compounded. The Ukraine, once known as the "Breadbasket of Europe," suffered famine at the beginning and was barely able to feed itself in the later years of the regime.

I witnessed a failure to generate bold and visionary research hypotheses in favour of safe work. At the Moscow Institute of Allergy, the staff had set up an extensive array of radio-assays to measure practically every hormone, and it seemed that all the available tests were being done on every patient, regardless of the complaint. When I asked about this unfocused protocol, they told me they were simply looking for correlations. I doubt I would have had the courage to do better if I too had had Stalin's repressions in my recent rear-view mirror.

What is to be learned from this historical vignette? Soviet biomedical and agricultural scientists were not acting with impure motives, but out of fear, and their scientific failures were grounded in their imposed ideology, not in a database. Lacking the freedom to test basic assumptions, they were limited in the hypotheses they could generate and the possible safe interpretations they could make of their data.

In those same years, the West tested many hypotheses and was able to forge ahead on a broad front, forming a robust understanding of the molecular scheme of life and based on that understanding, developing robust healing technologies. This approximately quintupled food production without increasing seeded acreages; not to mention the further benefit of gaining a humble sense of satisfaction from an integrated perspective of reality. Also in that interval, while the world population increased from two to six billion, large-scale famines outside of war zones were largely abolished from the face of the earth, while the USSR became a net importer of grains.

Intellectual integrity is an essential component of any successful long-term venture. Ideologies that provide *the answer* before the question has been fully formulated have no place in a progressive society. Stalin imposed communism as the essential framework of science in the USSR, and for that reason the scientists failed. He might have observed that other demagogues in history had similarly failed. Consider the imposition of a literalist interpretation of the Qur'an on a then-flourishing Islamic science by Hamid al-Ghazali in the twelfth century. This was a catastrophe from which the Middle East has not yet recovered eight centuries later. Nor is the West immune from this hazard; more than forty percent of Americans (and slightly fewer Canadians) want a literalist Biblical interpretation to apply to the teaching of science in the schools. Should their views ever come to prevail, the consequences for our own scientific progress would be equally disastrous. Assumptions are validated or invalidated solely by their outcomes; evidence alone can be the arbiter of what falls within the scope of truth.

A WORD IN YOUR EAR

There has been a new word added to the medical dictionary since I retired from medical practice seventeen years ago. In fact, there have been many words added, but the one I am thinking of is "theragnostic," because it encapsulates the essence of an imminent medical revolution. The concept is that there are drugs that have both diagnostic and therapeutic utility. In one application, radioactive molecules that bind to certain diseased cells were developed; these can be used first for the purpose of mapping out the extent of disease, and subsequently to treat it with the same drug is attached to a different radioisotope. The concept is powerful in the field of oncology, but it might also be applicable to some autoimmune diseases and infectious diseases that prove resistant to antibiotics. The idea has been around for a while, and it was a vision in the back of my head when I opted to concentrate my career in nuclear medicine more than fifty years ago. At that time, we had only one example of the principle at play, which was iodine in the settings of hyperthyroidism and thyroid cancer. Now it is becoming real in multiple ways because we have new tools for the job, and they work across a much broader range of diseases. I truly wish I could start my career again because the next fifty years will be productive beyond anything that preceded them.

Nevertheless, I have a problem with the word "theragnostic"—namely, that American journal editors want to reduce it to "theranostic," I understand why they want to do that, but it saddens me all the same because I think words should attempt to do justice to their antecedents.

The Gnostics (a big G word) were a small group of Greek thinkers who held that it was possible to use one's senses to build a data set; ie, to know and recognize that some things about the universe are true. They stood in contrast to others who averred that knowledge came via the whisperings of sundry spirits into the ears of susceptible and, supposedly, uniquely endowed people. The Gnostics emphasized the importance of personal experience in their belief systems and were treated as heretics; we might consider them the primordial forerunners of modern scientists. I would like to respect them for that because not everyone, even today, believes truth to be morally superior to a lie. The influence of the Gnostics tracks through our own language in words such as diagnosis (to understand the evidence), prognosis (the use of the available data to predict the future) and agnostic (admission of ignorance). Therefore, I am campaigning to keep the hard, glottal stop, despite its awkward phonetics, in theragnostic, knowing all the while that my cause is lost. As a passionate believer in the objective existence of truth, I want to avoid obscuring the path that our society took to break away from the oppression of witchcraft and sorcery to achieve an evidence-based awareness of what, who, and where we are. Some credit for this goes to the primordial Gnostics, and I don't want to blur them out of the picture altogether.

Linguists, who study the evolution of languages, report that the human mouth is intrinsically lazy. It requires more energy to pronounce a syllable that is voiced at the back of the throat than one that utilizes the front; accordingly, there are many historical examples of sound shifts toward the tip of the tongue being at play in the evolution of languages. In the case of "theragnostics," the first two syllables issue from the tip of the tongue on the teeth. Then "g" requires the word to move to the glottis near the voice box in the throat before the rest of the word again moves to the front. It's simply more efficient to first slur over the g and then to get rid of it altogether. The "g" will have to go; that's evolution, and I get it.

It is captivating to take note of how new words enter a language. Often, it involves the blending of existing words, but sometimes it's a simple case of importing the term from another language. While travelling in the Yemeni desert, we once made a rest stop and, while overhearing our drivers having an animated discussion in Arabic about something known only to them, I abruptly heard references to "automatic transmission" and "flat tire" bubbling to the top. It happened again when I was at the back of a doctors' room in a Yemeni hospital during morning rounds. Again, a lot of the discussion was incomprehensible to me until the Arabic was punctured by terms such as "emergency Caesarean section" and "blood transfusions." The Yemeni are only a generation away from the time when their society collided head on with the Western world; out of necessity, they have adopted much of the English technological vocabulary into their Arabic. Of course, the case of Arabic is rather unique, as there has been linguistic stability provided by the Qur'an, in that its words were supposedly written as dictated into Mohammed's ear and cannot be changed. English lacks a similar source of stability, and most of us have trouble puzzling out writings older than a few centuries because our language is continually sliding away from us as it ages; just try reading an eighteenth-century novel if you doubt my point!

I don't doubt that 'theragnostic' will have a short life, with or without the "g." It will evolve toward words that better describe variations on the technique; more new material will be added; the historical links will be progressively lost. There will be some interdisciplinary squabbling about which medical specialty is to claim responsibility for the business plan. Nevertheless, I remain very excited by the clinical promise of the new technology. Whatever it comes to be called; it promises a revolution toward hope.

ARE NEAT FREAKS
NECESSARILY MAD?

Mathematicians are the world's tidiest people; they use no unnecessary lines of text, every equation is balanced, and every proof incontrovertible. Oh sure, there are branches of derivative maths, like statistics, that deal with a range of wobbling uncertainties, but these exist only because the interface of mathematics with the real world necessitates measurements, which always carry forward a margin of error. And, just to be difficult, there is the field of quantum mechanics of which many prefer to say little except that it always gives the correct solution. ("Shut up and calculate."). Although science insists that evidence should be clothed in numerical garb, I don't think that pure mathematics itself is really science, so much as it is a distinct work of logic and a cognitive discipline. Mathematics has been, as Steven Weinberg said, "unreasonably successful" as a descriptor of the universe; nevertheless, mathematics and science remain distinct, though they are easy fellow travellers on the road toward understanding.

Mathematics seems to strive toward symmetries—the left side equals the right side, etc. Small wonder that it took until modern times for the realization to break through to the tidy disciplined minds of mathematicians that there are myriad settings in which even the short futures of deterministic phenomena become unpredictable. Isaac Newton might have had a premonition of Chaos theory when he refused to address the three-body problem in mathematical terms and stated that he expected God to rejig the universe as necessary to correct such gravitational perturbations as would be introduced by additional bodies in orbit around the Sun.

It is said that on a Sunday, when he found the sermon boring, Galileo's attention was gripped by a chandelier swaying in the breeze near an open window. Using his pulse as a chronometer, he determined the regularity of the pendulum and later confirmed that its period was a simple function of its length. By this means, he could predict the locations of the pendulum's tip for small angles of swing at any future time. It was as simple as $T=2\pi$.

Now pay attention to what happens if one hangs an additional pendulum from the end of the first; can the position of the second tip now be predicted? Not by the equations of classical mechanics. The rhythm of the pendulum has been lost; the motion is not that of randomness but of deterministic, yet unpredictable, chaos. What has become of the neatness of mathematical rhythms? We cannot be sure that Galileo ever came to this problem, but Newton did in the form of planetary orbits about the Sun, and he thought the problem insoluble. Had it been presented to Hamid al-Ghazali, he might well have thought it a demonstration of God's freedom to alter the rules of the universe as he saw fit from time to time. However, in modern times, chaos theory is understood as an expression of non-linear equations and nothing more.

Chaos is a basic theme of nature, and it recurs often in the profiles of landscapes—be they mountains on the horizon, incoming waves on waterfronts, the periodicity of tree branches and their diameters, or the configurations of flames on an open fire. I have a sense that the human mind needs this form of chaos as prophylaxis against madness. Why else would the market command premium prices for waterfront cottages or those with a grand view of a larger natural scene? Why is a bonfire or fireplace such a relaxing component of a great evening? Why is a walk in the park an effective way to reduce an elevated blood pressure—better even than a walk along the sidewalk? We are learning that a daily walk in a natural environment is an effective prophylactic against dementia; does a chaotic scene replete with trees, squirrels, and barking dogs play a role in this? Does music, with its own forms of chaos and variations on a theme fall into this as well? The short answer is that we don't know the answers to these questions yet. I suspect that there is more to chaos and our mental health than just relaxation; there are likely intrinsic chaotic rhythms resonating naturally with our neural networks that need us to break away from rigid, imposed concrete geometries and spreadsheets from time to time.

A MEDITATION ON THE PERIODIC TABLE

When I woke up this morning, it occurred to me that I must write about iodine. Iodine resonates with a long-standing theme of my medical practice: thyroid glands. Moreover, iodine's impact on wellbeing has important historical and socio-political implications.

Iodine is the fifty-third element of the periodic table (ie, the alphabet of matter), as it was laid out for us by Dmitri Mendeleev in 1869. At that time, the element that would eventually be known as iodine had not yet been discovered, but Mendeleev left an open spot in the correct place on the table and accurately predicted its properties. In subsequent years, others demonstrated that iodine plays an important role in human health. One of its most obvious contributions is preventing the formation of benign lumps in the thyroid gland, otherwise known as goitres.

We've known about this for over a century. The regions surrounding the Great Lakes were once known as the Goitre Belt of North America because those agricultural lands are deficient in iodine. School class pictures of the 1930s show visible goitres on about a third of the girls. When I started practicing medicine, women from the Bruce Peninsula still came to our clinic with iodine-deficiency goitres because they were scooping non-iodized salt from the cow barn for their kitchens. In Canada, we passed a law stating that all salt sold for human consumption must be iodized.

The Americans, by virtue of an accidental choice of words in their Constitution, remain free to grow goitres if they so wish.

Over time, we have learned that iodine is an essential component of thyroid hormones, which have many roles throughout life but are especially critical for brain development during pregnancy and infancy. The World Health Organization (WHO) recognizes this fact and notes that despite many efforts, iodine deficiency remains a persistent problem for a third of humanity and amongst affected populations, IQ scores are three–five points lower.

Unfortunately, some of the WHO's touted success in overcoming the deficiency has been merely cosmetic. I saw this in Khartoum, Sudan, where the question of endemic iodine deficiency was discussed with a senior WHO official, who assured me that the problem had been dealt with. However, upon further inquiry, it turned out that while the Government had passed legislation requiring all grocery stores to stock iodized salt, there had been no public education concerning its benefits. The grocers grasped that this type of salt had been in some way manipulated, so they marked up the price to reflect its increased (yet speculative) value. Customers, who were always looking for a bargain and knowing nothing about iodine deficiency, avoided it because of the price. The situation equilibrated when every vendor kept one box of iodized salt on the top shelf to be compliant with the law and stay in business, but nothing else had been changed.

One impact of the Chernobyl nuclear reactor accident was that many nearby children, who had been exposed to radioactive iodine, subsequently developed thyroid cancer. The Chernobyl region is chronically iodine deficient, and iodine supplements were provided to the population in the early days of the Soviet regime. At some later time, the Party announced that the deficiency had been resolved, and the supplements were discontinued. At the time of the nuclear accident, deficiency was once again the norm. Consequently, residents' thyroid glands had hugely increased their uptake of iodine, and they did not discriminate between the stable form and the radioactive isotopes. This was especially true of children below the age of five and most of the later cancers came from this cohort. If the Chernobyl accident had occurred in Canada, it would have had catastrophic socio-economic effects, but it would not have produced an equivalent number of cancers, because our diet contains ample amounts of iodine, and the uptake of radioactive iodine would have been a small fraction of what the children living in the Chernobyl region actually received.

There is a long history of goitres along the banks of the Nile River. The Egyptian Pharaohs maintained that they were divine, and this grounded their supposed right to rule over ordinary people. But the justification created a problem: how are gods to reproduce themselves without polluting their divine gene pool? These supposed deities believed their myth so much that they resorted to incest to maintain their pure bloodline. This strategy had serious deleterious effects, one of which was generations of children with congenital goitres in the royal family—not because iodine was lacking in the diet, but because a mutation had inactivated the enzyme that adds iodine to tyrosine to make thyroid hormones. Since goitres occurred often in these inbred "children of gods," they were promoted to the population as a feature of female beauty. Cleopatra, the last Pharoah, had a goitre and proudly promoted the fact by ensuring its appearance in her image on the Egyptian coinage. A local doctor from Sudan told me that to this day, men along the Nile villages prefer a woman with a generous goitre; it's just one extra curve on the female silhouette—what's not to love about that!

Goitres also play a role in art history. Artemisia Gentileschi's sixteenth century painting of the beheading of the Assyrian general, Holofernes, depicts Judith and her maid making their escape from the scene, but turning in a vital moment when they hear a noise behind them. In this dramatic moment, the artist focussed the light on Judith's fulsome goitre. Might it be that the artist thought of goitres as an accessory or beauty feature? Or might it have been the case that the artist's model had a goitre?

The periodic table is a marvelous construct that has great stories stuck into every one of its crannies. Like a lot of science, the stories typically begin by being simply true, and then beautiful. Then, surprisingly often, they slip into becoming useful as well. That's how we got modern chemistry—warts, goitres, and all.

A TRIP TO THE ZOO OF SUBATOMIC PARTICLES

In fifth grade, my teacher taught us that there are three states of matter: solids, liquids, and gases. For many of us, that was enough science for a lifetime, unless one thought to ask questions like; "what is the smallest thing in the world?"; "why can't we see through a candle flame?"; etc. Most of us didn't bother to ask much.

At the same time, in a place far away from my classroom, there was a troublesome astronomer named Fritz Zwicky, who held a research job at the California Institute of Technology. Zwicky was, by all accounts, a disagreeable man, who fought with colleagues and disrupted meetings with insults and name-calling. But he was blazingly brilliant. He measured the rotational velocities of stars around the galaxy and determined that the amount of matter in these systems was insufficient to hold them together; yet hold together they did. Unless there was more matter in the galaxy than the solids, liquids, and gases that could be measured, the stars would be flung out by centrifugal forces, like pellets from a slingshot. Zwicky postulated "dark matter" (i.e., matter that was undetectable with telescopes) to solve this puzzle. But this postulate didn't gain traction at the time because of his combative personality. A generation later, Vera C. Rubin, one of the world's first women astronomers, and a much-loved science personality throughout her long life, undertook a careful study of galactic rotations. She confirmed that the mass of dark matter exceeds that of normal matter by a factor of six.

We now know that every galaxy is surrounded by a cocoon of this mysterious substance. Without it, the universe would be a disordered shooting gallery of matter in random motion. We call it "dark matter" because we don't know what it is, but we know it is there on account of its gravitational attraction. Dark matter does not emit any electromagnetic signals (e.g., radio, light, infra-red, or ionizing radiations) by which we might characterize it. The theoreticians tell us that it is likely a sub-atomic particle of some kind, but the Large Hadron Collider (LHC), the world's most powerful atom smasher has so far, not detected, it despite many attempts.

For the present, we are stalled on the nature of this unknown. So, let's look at another problem that was just as mysterious and elusive in its time, but that was eventually solved by scientific persistence: the story of neutrinos.

In the early years of the twentieth century, physics was in an intellectual crisis. The electromagnetic spectrum was under intense scrutiny as diverse forms of ionizing radiations were being detected and classified. Ernest Rutherford clarified the matter of alpha radiations, which are simply energetic helium nuclei—atomic cannonballs, if you like, that knock over whatever they impact. The gamma rays are very different, consisting of pure energy, as identified by Henri Becquerel. Both alpha and gamma radiations have specific energies of emission that fit exactly into the atomic theory equations. But the real problem came from the beta radiations, which do not emit the correct energy required to balance the equations. That imbalance threatened the very foundations of theoretical physics.

A solution of sorts was offered by Wolfgang Pauli, who suggested that an as-yet undiscovered particle with no electric charge carried away the missing energy. His proposal was eagerly embraced because no one had a better idea to offer. But despite its popularity in the scientific community, Pauli himself was consumed by doubt and shame for all his life because his hypothesis remained unproven. In his private diary, he wrote something like this (in German):

> *What have I done? If I had suggested that the energy is carried off by the tooth fairy, people would be shaking their heads and suggesting that I should see a doctor, but because the imaginary particle has been given an exotic name, they say I solved a problem.*

The mystery was resolved a year after Pauli's death, and the particle is now called the neutrino (in Italian, the "little neutral one"). We still do not know the exact mass of neutrinos because they are smaller than all measurement scales, but we do deduce that their total mass is greater than the mass of all normal matter in the universe. We also know that they have three mass/energy levels. Dr. Arthur McDonald from Queens University resolved this problem with an apparatus he built at the bottom of the Sudbury nickel mine and, for that he received a Nobel Prize.

Every second, there are millions of neutrinos passing through every square centimetre of your body; they will go on doing that, undetectably, for as long as you live. Perhaps a few times in your life, there will be a head-to-head collision of a neutrino with the nucleus of one of your atoms, and there will be a flash of light that no one will see. But otherwise, there will be no consequence. We can't avoid them, because neutrinos easily pass through the Earth. Tens of millions of kilometres of lead would be required to reduce their flux by only about half.

For the most part, the neutrinos that we experience come from the centre of the Sun, where they are created when hydrogen is fused into helium. They are also produced during beta decay of radioactive isotopes, some of which are used in treatment of medical conditions like thyroid and neuroendocrine cancers. I happen to be very pleased with neutrinos, because they tell us that the Sun will continue to shine for another 5 billion years, and they are also the major reason

I had a job in a hospital for most of my career. Pauli would have been so pleased, and he would have used the example of the elusive neutrinos to encourage us to keep on trying to understand dark matter as well.

I'm amazed every day that the zoo of atomic particles—of which dark matter and neutrinos are but two—is out there and preparing to bring us another day on Earth. What is this place we call the universe? I don't suppose that my primary school teacher had a vision of how she was bending the minds of her charges; perhaps teaching is like spraying the classroom with neutrinos: most moments are too small to matter but, taken in their totality, they substantiate a universe.

CODEBREAKING

When I was an undergraduate, physicists and chemists loved to tease biologists. "Look," they might say over a drink, "we pose our questions and theories in quantitative terms, and we can write predictive hypotheses that direct the experiments we can perform. You biologists are mere stamp collectors who debate over the relatedness of teeth and bone fragments someone found in a swamp. Your field seems not to have had a new idea since Charles Darwin." At that time, no one noticed that things were changing. But change they did, and they are still changing. The production of Covid-19 vaccines in just a few months is evidence that the era of biologists amounting to obsessive collectors has long since passed, and in this era biology has predictive power.

Since about the mid-1900s, biology research increasingly became about codebreaking, as it became obvious that life follows unifying sets of rules. It began with James Watson and Frances Crick's discovery of the structure of DNA: four bases arranged in pairs along two intertwined strands, always A opposite T, and always G opposite C, making for 3.2 billion iterations along our forty-six chromosomes. and no two humans are ever exactly alike.

If this looks like a digital computer code, that's because it is. The code was broken in 1951. Next came the challenge of reading the sequences of the individual A, T, C and G sequences along the DNA. It required a singularly expensive research initiative—about ten billion dollars (in current dollars), which was funded by the American Congress over a decade—to map out the DNA sequence of just one individual. About a decade ago, Congress asked for an accounting of the benefits that had materialized from this expenditure and found a return of $141 for every dollar spent on the initiative—and still counting!

During the days of the initial sequencing, I attended a medical conference at which the then-forbiddingly high costs of DNA sequencing came up. I pointed out that costs were falling; they were already down to one hundred thousand dollars, from the original ten billion dollars. Moreover, the new protocols took less than a week to complete. I was interrupted by another researcher who mopped up the floor with me over my unseemly optimism about the future affordability of the technique. But today, a limited DNA sequence costs less than a hundred dollars—a hundred-million-fold price reduction. Tracing lineages with DNA has become a weekend parlor game that every family can afford, and that no police force can do without.

It was known from prior research that DNA somehow coded for protein structures. But how can a four-letter DNA code for a twenty-letter amino acid sequence? That remained a mystery until researchers stood back from the problem and asked, "What is the RNA about?" RNA (ribonucleic acid) bears a resemblance to DNA that I won't detail here. It is also a four-letter code, but its polymers are single, rather than double-stranded, in form. One kind of RNA turned out to be a long set of polymers (aka "messenger" or mRNA); mRNA turned out to be templated by DNA in the chromosomal genes. That, in turn, was the template for the proteins required to maintain the structure and function of the cell. mRNA is the inter-office memo in which the executive office tells the floor manager what the market requires today. This insight into mRNA resolved another mysterious code.

Still, there remained a gap to be filled. There are other small RNA molecules whose role became clear when it was observed that a unique molecule was responsible for the transport of each of the twenty amino acids; they are collectively known as transfer RNA (tRNA) because they transferred each amino acid to its designated location along the mRNA, not unlike ushers who conduct guests to their assigned seats at weddings. Each step forward solved another puzzle.

Finally, there is a class of large RNA molecules that fold up to create a structure known as a microsome, and this structure accommodates the mRNA and the tRNAs as they deliver the amino acids. The complex acts somewhat like a zipper to build the protein. That was another step in the process; codes layered on codes, but all had proven to be soluble.

By now, biology seemed to have become discouragingly complex. But simplification lay ahead, as it was demonstrated that all life on Earth—microbial, plant and animal—operates with the same genetic code, and all manifestations of life are demonstrably related. No longer do biologists classify specimens like stamp collectors with heavy reliance on external features. It is easier, and less costly, to determine a DNA sequence than to do anything else. From slime molds, to cabbages, to humans, all have the same genes and scattered mutations to account for our unique differences. In short, modern biology has identified its Lego blocks, and now we can mix-and-match or delete the components to specific ends, such as in the production of vaccines. That is also how *E. coli* bacteria fitted with the human insulin gene now produce the world's supply of safer insulin. These techniques were also used by plant breeders to multiply the yields of food crops over the past seventy years. Canadian fisheries industry scientists used molecular biology to develop a strain of fast-growing salmon that give us affordable fish to eat, and so much more.

The mRNA-based Covid-19 vaccines have been criticized in some quarters because they were produced too quickly. "Somebody's cheating," it was alleged. It is true that all previous vaccines required years of research to achieve the desired product. The allegations about the mRNA Covid vaccine overlooked the advances that basic biology made in the past 70 years, and the knowledge base that the vaccine producers now have available. These resources allowed investigators to determine the molecular structure of an anti-Covid vaccine and how it should be manufactured in only a few days. After that, it simply was a case of ordering the necessary components from biology research supply companies and assembling the vaccine. This was done in a matter of only weeks, and the rest of the time before the distribution was devoted to assessing issues of stability, safety, and efficacy. The pandemic was the first test of the new paradigm, and it came through with flying colours. I notice now that new vaccines are produced on short notice as mutants of the virus arise. There were no shortcuts taken with the mRNA vaccines; this was simply modern plug-and-play biology at work, and this will be the way of the future for the production of vaccines and other biological products for medicine.

Indeed, on January 22, 2024, the World Health Association launched a program to vaccinate children in Cameroon against malaria, expecting to reduce the mortality of this childhood disease by a third. Scientists previously spent a century attempting to produce an anti-malarial vaccine without success. The new product is an mRNA vaccine that owes its success to learnings from the Covid-19 experience.

We can only cope with this hierarchy of biological codes because computers have been evolving about as quickly as the biology. Molecular biology research would be impossible without large computers to keep the codes sorted. Much of the Allies' success in WWII came about because the German military's Enigma codes were deciphered by Alan Turing and others, with early versions of computers. This decoding work has been estimated to have shortened the war by two years and may have spared up to fourteen million lives. No one in that historical episode accused anyone else of cheating. I find myself thinking that the codebreaking mindset that achieved Enigma decoding transferred into biology when the same people transitioned from military to civilian lives in the post-war period. Some went on to academic biology, and they brought that new approach along with them. Let's give credit to honest codebreakers; they're all on our side.

I don't intend to demean the importance of people who continue to collect fossils and other rare specimens; Tiktaalik fossils and the like will always cause my pulse to race. But DNA is a stake, written in the language of nucleic acids, that allows a closer look at the relatedness of life, and it buttresses the fossil stories as well. Those old WWII stories about code breakers also made my heart race, as does the story written in DNA today.

THE VANISHING STETHOSCOPE

When we are ill, it is comforting to know that our clinics and hospitals are well equipped with a wide array of tools that can elucidate the meaning of our discomfort. It may surprise you to learn that prior to 1816 (a mere two centuries ago), medicine had no diagnostic tools. The first tool in diagnostic medicine was the stethoscope, and its story is something like this.

There are several anecdotes about how Dr. René Laennec supposedly came upon the principle of the stethoscope. In those days, auscultating the chest required the physician to apply his ear directly and firmly onto the patient to hear their breathing and heart action. One day, a patient with an offensively diseased chest came to Dr. Laennec for a consultation. All the required steps in the physician/patient protocol had been followed: a history had been taken, the patient had been inspected, and all the steps of the physical examination that required direct contact with the patient (ie, observation, palpation, and chest wall percussion), had been completed. The next step would have been to listen to the heart and lung sounds. But with every fibre in his body, the doctor did not want to press his ear to this patient's chest. He paused and looked around his office. As he did so, a memory from childhood flashed through his mind: he and his brother had listened to their parents' arguments through the wall of their room by pressing their ears onto cardboard tubes held against the wall. In a flash of insight, Laennec reached for a sheaf of papers on his desk, rolled them up into a tube, and applied one end of the tube to the patient and the other to his ear. He was amazed that he heard the breath and heart sounds better than he had ever heard them before.

As with any invention, many—including Laennec himself—attempted to improve on the prototype. Laennec determined that the best resonances were obtained through a hollow wooden tube of about twenty-five centimetres in length. Many other investigators followed over the next decades. They debated the fine details of the best stethoscope. For instance, was an endpiece fitted with a diaphragm more sensitive than one with a bell-shaped end? For which applications? Seventy-five years later, in 1891, there were still arguments published in the medical journals about the acoustic qualities of stethoscope bells turned from walnut versus cherry wood. For patients who could afford to pay the extra fee, there were also stethoscopes fitted with ivory bells! As recently as 1958, when I was a first-year medical student, my teachers thought it worth their time to instruct us regarding the optimal firmness of the rubber tubing to look out for when we purchased our own stethoscopes.

In those early years, there were several books published on the new art of stethoscopy, in which the authors argued for recognition of this new skill as a medical subspecialty requiring mandatory training for its would-be practitioners. It was not unlike the arguments I recall from the seventies and eighties over the emergence of office-sized ultrasound machines. Who would be qualified to use them? And, more specifically, who was entitled to charge an extra fee for their deployment?

Today, one can purchase a so-called butterfly stethoscope, which is actually an ultrasound sensor that plugs into a cell phone such that the examination can be recorded. This includes images and the dynamic blood flows that are responsible for heart sounds. These stethoscopes also allow for play-back or transmission to an expert for a second opinion.

At this time, you can purchase an electrocardiograph for your own use and transmit signals directly to the physician's office for immediate interpretation of an irregular heart rhythm. It's not quite a Star Trek tricorder yet, but the resemblance is remarkable.

WHO FARTED IN THE ELEVATOR?

The most urgent question in all of science right now concerns the matter of the origin of life on Earth and its many corollary mysteries. Are there other examples of life or its analogues on other worlds? It's always been a source of intense speculation, but the interest level has amped up in our time because these questions might become answerable in the near future. This is in part due to the new astronomical tools that have come online in this decade with more yet to come.

The idea of life on other worlds is not new; it was hypothesized more than four hundred years ago by Giordano Brno, who speculated that the stars might be distant suns with their own planetary systems. For this suggestion, he was condemned as a heretic and burned at the stake in the city square of Rome in the year 1600.

Now, in our own lifetime, we have verified with the Kepler telescope that every star has one or more planets circling it, and there are many more planets in our sky than there are stars. Many exoplanets show no prospect of being habitable, even from this distance, but about one in five planets may be of the right size and have both liquid water and an atmosphere.

The next question: do the conditions on any of these planets permit even the possibility of life as we know it? Prospects are good that there are some that do. But for now, we have only one example of a planet that harbours life, and that planet is Earth. If we want to have a shot at recognizing life somewhere out there, we must train for it by recognizing all the life that exists around us in the here and now—a task we have not yet completed. To go one step further, intelligent life on an exoplanet is an even longer shot, but it is not an impossibility.

In Brno's time, the only life forms recognized on Earth were the macroscopic plants and animals. However, in his era, a Dutchman named Anton Leuwenhoek, taught himself to grind magnifying lenses with which he observed that the detritus he scraped off his teeth, and out of his other orifices, was teeming with things that lived and moved. He called them "animalcules" and thus opened a window on life extending into the microscopic range.

More than sixty years ago, when I first studied medical microbiology, we thought that we had all those tiny life forms identified, named, and catalogued based on their shapes, staining properties, food preferences, the diseases they caused, etc. However, soon after that, we found that only about one percent of all bacteria could be identified in these ways. Most species refuse to grow in pure cultures in laboratories without the collaboration of their natural neighbours.

In addition, we had completely overlooked an entire swath of living things that are also small but are not bacteria; these are now known as archaea. They are as numerous and as diverse in their nature and distribution as the bacteria. Archaea live everywhere; we have never dug a hole so deep into the Earth that they weren't waiting for us at the bottom. Every microscopic crack in every mountain rock and every hot spring on the ocean floor contains archaea. Every handful of garden soil contains tens of thousands of species of archaea and bacteria. It is tempting to speculate that some of these might guide us to the origin of life via the study of their comparably simpler chemistry.

There is now a small field of speculation that deals with concepts of pre-biotic evolution. The field involves such things as the spontaneous ordering of molecules into a resemblance of enzymes on the surfaces of rocks that lead toward the earliest biochemistry of archaea and/ or bacteria. If I had time to get into the field and build a framework for study, I would begin with geology and thermodynamics before bridging to questions of biological evolution. We still have so much to learn about life on Earth, never mind the other planets! Yet, here they are, almost within reach, whether we are ready for them or not.

If there is life on other planets, how will we recognize it? Don't look for "Welcome Earthlings" banners floating from balloons as our probes descend from their sky. For starters, from a safe distance; ie, from the Earth, we will use a technique called spectroscopy to look for prerequisites of life as we know them and atmospheric changes, possibly mediated by life in the atmosphere of candidate planets. Indicators like oxygen, carbon dioxide, carbon monoxide, water, methane, and industrial pollutants will provide unique signatures in those spectra obtained from space telescopes. Certain mixtures will give away their biological origin. I think of it as the "Who farted in the elevator?" approach to the search for exotic life. I seriously doubt that we will encounter highly evolved thinking creatures waiting to welcome us when the doors open on the next floor, so to speak. If we only identify gaseous products of metabolism out there, it will suffice to overturn some of our most deeply held historic ideas about the uniqueness of life on Earth and our place in the universe. I want to know how this will work out over the next decades. Through the smoky dust, past the smell of burned flesh and the dimming of time, I can imagine the vindicated grin on Giordano Brno's face.

KEYSTONE ARCHES

I first encountered the concept of "keystones" in ecology when I was preparing for a safari in south and central Africa. I had read Sean B. Carroll's book *Rules of the Serengeti.* The problem he described concerned the viral animal disease called "rinderpest," the name being derived from Dutch for "cattle disease." The epidemic was rapidly wiping out not only the imported domestic herds, but also the treasured large wild mammals that are unique to the African savannah.

As if that was not a serious enough matter, at the same time, native species, including many non-ruminants that are not subject to rinderpest, were also diminishing sharply in numbers. The wildebeest seemed to be especially at risk. as they were dying in disproportionately large numbers, and they were assumed to be the primary reservoir of infection. It seemed unlikely that a practicable method could be found to immunize all the wildebeest, and a decision was made to attempt to at least save the domestic cattle. The vaccine was developed, and the cattle were immunized with great success, whereupon the disease also disappeared from the wild. Other species of wild ruminants also recovered and began rapidly to reapproximate their former numbers. Surprisingly, even the species that were not prone to rinderpest then experienced a sharp recovery toward their former numbers. Several explanations emerged from the experience. First, the viral reservoir was never in the wildebeest, but in the domestic herds and the disease abated once those had been immunized. There followed then a series of second and third order significant events. The increase in wild stock, especially of the wildebeest, increased the extent to which the soil of the savannah was once again turned over by animal hooves in the annual migrations, thereby mixing their manures into the soil and increasing the permeability of the soil to rainwater. The new growth of grasses was more abundant. The herds of wildebeest served in this series to stabilize an ecological arch.

I'll offer a couple of additional examples of keystone effects operating a little closer to home.

Wolves were eliminated from Yellowstone Park many decades ago because ranchers deemed them to be predatory on their herds. In the absence of wolves, the elk proliferated and, over time, became relaxed about lounging near the streams where they came to drink. They forgot that wolves might lie in ambush there. In turn, they browsed the poplar trees along the waterways to a much greater extent. In the absence of trees at the water's edge, soil erosion became an issue, and silt began to dam up the streams, causing flooding in the grassy plains that included places where domestic animals also grazed. The reintroduction of wolves encouraged the elk to promptly return to the uplands after a drink and not content themselves with picnicking along the water's edge. It has been a success all around, and even the ranchers, who were once opposed to the plan, are now satisfied that a mutually beneficial balance between restoration of a natural environment and an economically successful industry has been achieved.

My third example comes from my own experience of living for thirty years in a forest in southern Ontario and volunteering with expert ecologists at the Thames Talbot Land Trust. I mainly worked to recover a forest plot that had previously been damaged by inappropriate agricultural exploitation. The primeval forests of southwestern Ontario were once heavily populated by oak trees, many of which were removed for lumber income. Oaks need generous sunlight to grow. They gain a slight survival advantage over competing rapid growers in nature by surviving small forest fires and eventually achieving a position in the forest crown. From that position, mature oaks come to tower above the forest, and their leaves provide food for many species of insects. In turn, these insects become a rich protein smorgasbord for birds. Even hummingbirds need dietary protein to raise their young.

While we still have some large oaks in southern Ontario, there are possibly not enough small trees to take the place of mature trees as they are harvested. The problem seems to be that some alien grasses are proving to be toxic to young trees. Proteins require a supply of all twenty amino acids for their synthesis in plants, as well as in animals. Plants are skilled at chemical warfare, and grasses that also need access to sunlight have, in some species, evolved to make a form of the amino acid tyrosine (meta-tyrosine) that is toxic to trees in this warfare over access to sunshine. The fescue grasses are alien species used to provide cover for golf courses and horse pastures, and who have slipped under fences into adjacent woodlots on more than a few occasions. There, they are now a problem for the ongoing regeneration of oaks. There is very little grass in my former woodlot, and there are now quite a number of young oak trees— more than a hundred currently line the trails and the driveway. I did also sometimes cut back competing growth to admit light, and I have high hopes for the state of the driveway and the trails in the future. A healthy stand of trees will also help to ensure the health of my friendly hummingbirds at the house. Clearly, the oaks are a local keystone that, in this case, is stabilizing the recovering forest.

LIFE IN THE SLOW LANE

It's a slow weekend in the middle of the Omicron wave of the Covid-19 pandemic, and the residents of my condo building are following sound advice to isolate. You could safely fire cannons down the corridors and not hurt anyone. I've been in self-isolation since New Year's Eve, warding off the upsweep of the pestilence. Today my oldest granddaughter tested positive. I don't think I am at high risk of contracting the virus, but I am being careful. So, I'm looking at about another week of isolation, and I need to find something outside of myself to complain about.

I was so pleased to be able to watch the launching, and then the final unfurling, of the James Webb Space Telescope (JWST). The rest of the setup and calibration will follow over the next six months. They won't be quite as heart-stoppingly wonderful on a moment-by-moment basis, but they will still be of interest. As the eighteen telescopic panels are brought into a unified focus, signals are calibrated, and space-to-ground communications are tuned. Within the coming months, we expect to look back at approximately the first two billion years since the Big Bang and watch the first stars in process of formation—you could say, in real time! Lest you wonder how this works, remember that light takes time to travel and what we see represents the star in question as it was when that light was emitted, as long as 11 or 12 billion years ago.

The thought reminds me of my experiences with Royal Festival Hall concerts in London, England, nearly sixty years ago. We could only afford the backrow balcony seats, and the music came late to our location in the theatre. In the present instance, JWST is providing us with the only available—albeit backrow seats—and still, we will see the actual act of creation on our newly extended, nearly thirteen-billion-year horizon. It may be a while longer before the data fully make sense, or before we get over our initial giddiness and become comfortable with creation's expanded lightshow.

In a lockdown, there are other, equally giddy things to consider, even without a telescope. I have no patience for the classical philosophies of the Aristotelian, Thomist, or Augustinian variety; these were people who were not inhibited from writing down their reflections in multiple volumes, even though they couldn't distinguish between divine inspiration and dyspepsia. For many years, I believed that my failure to comprehend these and other august figures was my fault—that if I only took the time to listen and think more carefully, my blindfolds would fall away and all would be well.

So, I took a philosophy course a few years ago, hoping to make a breakthrough. I endured many hours of intellectual bullying before it came to me three lectures from the end: the problem was not mine. There simply is no *there* there. I walked away, and I've felt so much better ever since.

I've recently come upon a handful of modern philosophers, like the late Daniel Dennett, who express their philosophy in terms that begin with objective data. They consider what we know about the fabric of space-time, the working of evolution, and brain functions. For the first time, I suspect that philosophy, too, must learn how to take instructions from the universe as a prerequisite to telling us what is true. We, ourselves, are part of the truth. Stop looking for hidden meanings in reinterpreted mutterings from the past and take note of where society is now. It's not that we don't need philosophy anymore; we just don't need those outdated pronouncements anymore.

What we do need is philosophers who acknowledge that there are real truths and real values to ground us in what we know from evidence. Truth does not come to us as a prefab, fat-free package that explains itself. Before there can be hope of an answer, it is up to us to formulate the answerable questions, and that will be the ongoing work of generations. It can't just be any question that we happen to want to be true—it must be the questions that will lead to truthful answers, whether we want them or not. I'm looking to JWST and its family of other new research technologies to provide the basis from which to refresh philosophical discussions about humanity's place in space, time, and matter.

TRUE STORY

The Middle Ages were times when would-be artistic performers were developing the craft of what became rock concerts about six hundred years later. Troubadours roamed the European countryside strumming on rudely made lutes and singing songs of unfulfilled love, hoping to attract a free meal and a place to sleep inside the castle where one could secure one's back against a wall, which was about as safe as was possible in those times. It would be a bonus if one tumbled into a furtive, dark, nocturnal intimacy that could fuel fantasies to inspire the next day's song.

One such wandering minstrel of storied fame was trying to catch the eye of an especially beautiful maiden, who remained on the periphery of his performance. The more she kept to a safe social distance, the more desirable she seemed to him. Eventually, he managed to corner her with a proposition of eternal love, to which she replied, "Good Sir, why do you want to make love to a corpse?" Then she ripped open her bodice to reveal a large odorous ulcerated cancer on her breast. In response, the stunned young man fell his to his knees, confessed his sins, and took a vow of celibacy on the spot. Thereafter, he supposedly redirected his life to support needy women. That's the story, and I believe the part about the cancer.

At the cancer hospital in Sanaa, Yemen, we encountered neglected cancers of many kinds with some frequency. I can't forget the women presenting with ulcerated and infected breast cancers.

Generally, it was the smell that gave them away, even while they were still the length of the hospital corridor away. Those suffering women, chained to their own dying bodies, were reminded with every breath that death was stalking them. And it wasn't always quick demise, unless the ulcer mercifully became the conduit for a rampant infection. I recall a woman, so afflicted, who told us that she didn't want to die alone, but the odour was keeping her family away, and they had moved her bed to an out-building. She said, "No one can bear to be near me." We suggested that dressings impregnated with charcoal might be helpful to absorb the smell, and we hoped they eased her dying. I had forgotten until then, but have never forgotten since, that in the long arc of human history preceding modern breast cancer surgery, what I witnessed there had not been a rare event.

In 1889, an industrial benefactor granted funding for the construction of the Johns Hopkins Hospital in Baltimore. Dr. William Osler, originally from Toronto, was invited to assume the post of physician-in-chief. Dr. Osler's first task was to select a cadre of skilled physicians for the hospital's departmental leadership posts. For the Department of Surgery, he chose William Halsted. Halsted had been considered one of the most brilliant young surgeons in New York, but he had accidentally become addicted to cocaine while using himself as a subject in pain control research. As addiction took over his life, he dropped out of sight and became a derelict. Osler found Halsted while he was recovering from a stay in a mental hospital and convinced him he was needed as chief surgeon at Johns Hopkins. I can't think of a more unlikely second chance, but it shows that Osler was an excellent judge of character.

In his new post, Halsted commenced an in-depth, career-long study of breast cancer, especially of how it spread through the body from the origin. From his observations, he developed an operation, known as the radical mastectomy, which became, and long remained, the standard treatment for primary breast malignancy. It certainly saved many thousands of lives, but it also had many detractors. Feminists about fifty years ago were outspoken critics of the operation—and of Halsted himself. They declared that the operation required to be tailored to individual variations to avoid the disabilities and the defeminization of the procedure. Having witnessed the debility induced by radical mastectomy in my own family, I know what sparked their outrage.

In defense of Halsted, I will say that he was the first surgeon to approach breast cancer with an evidence-based mindset. That said, with only limited data at hand, he made mistakes, and it is the role of pioneers to pay the penalties for having been first. His modern critics have forgotten how dreadful an ulcerated and infected cancer was for many patients and for families through history. His operation was a step forward, and we should be thankful for his historical contribution while acknowledging that the frontier has since shifted, and we have moved on. In current practice, only the wedge of tissue containing the tumour is resected—not the entire breast, unless the stage of the disease demands it. Moreover, about three decades ago, it became possible to image the lymphatic drainage pathways of each individual cancer (lymphoscintigraphy), and that procedure now directs the surgeon's attention to a few suspicious lymph nodes. Thus, only the primary draining nodes are removed while sparing as many as possible and avoiding the debilitating arm swelling of the past.

Today, except in places like Yemen, the breast cancer frontier is less concerned with large tumours than the behaviour of any tumours, including the smallest, which are only detectable with good ultrasound and accurate needle biopsies. Now, our visionary goal is to avoid unnecessary interventions of every kind. For the past 170 years or so, there has been no way to argue against a cancer diagnosis if it was made by an experienced pathologist equipped with a good microscope. That is still a good technique, but it may be about to change. We are learning that there are sometimes lesions that look like cancers under the microscope but that may never develop into a disease that needs treatment. To make this distinction, we need new tools that surpass the microscope.

I am looking to my own field of nuclear medicine, where radioactively labelled drugs that selectively seek out cancer cells can be used for both diagnostic imaging and therapy in the new field of theragnostics. There are realistic prospects that, in the future, some such drugs will make better distinctions between cancers and non-cancers than a microscope, and it may become possible to treat them with this class of agents without surgery. It won't happen soon, but it is becoming possible to dream about it. The validity of the technique has already been demonstrated with some other cancers. When that reality comes to pass, we will ask the minstrel to sing a different song.

A MESSAGE FROM THE IVORY TOWER

Writing for scientific publication is, at its best, a team sport. Authors are reasonably comparable to quarterbacks, whose roles support a strategic play. Unless it is properly set up, the manuscript has small chance of going forward successfully. If you've ever written for publication or a committee or on a corporate assignment, you will have experienced the intervening referees and editors who stand between your work and its ultimate purpose. It can be a long road. It can even take years to ensure the documentation of the details, down to the last comma. Despite the authors' frustrations with publication delays, in most cases, the critics are not there to block, but to improve insightful writing to best penetrate the minds of readers. Their role is both to inform and to persuade.

It's not surprising that nature also uses an array of modifiers to ensure that life's messages are interpreted appropriately by the prevailing environment. Just as we thought that we had unambiguously decoded nature's DNA "Book of Life," it turns out that the environment in which life finds itself has options for overriding the seemingly rigid genetic sequences that were laid down in DNA.

Consider the example of the Dutch Hunger Winter. In early 1945, the Germans still controlled the Netherlands. When food became scarce, they decided to divert what resources they still had to their military. The Dutch population began to be hungry, and more than twenty thousand people died of starvation before the Allies liberated the region. The Netherlands had not experienced as much fighting or infrastructure destruction as some other parts of Europe, and some aspects of some peoples' lives had continued in a quasi-normal fashion throughout the war. Specifically, pregnant women continued to make regular visits to their doctors, and their conditions were carefully recorded. Women who delivered under starvation conditions had underweight babies, but when these babies were adequately fed after the war, they quickly caught up and seemed to grow normally thereafter. In adult life, these individuals tended toward obesity, and many developed type 2 diabetes and its complications. One could say that the ever-vigilant environmental editor had sent a message to the fetus in utero saying, "It's a tough world out there, kid. You'll want to hang on to every resource that comes your way. If you don't need that bite of food right now, eat it anyway; put it in your fat stores, and damn the consequences, because life may otherwise be short." There were no mutations involved, but that editorial note opened the doorway to type 2 diabetes for many in decades to come.

From continued follow-up of that Dutch cohort, we learned that type 2 diabetes appeared again in the next generation; this was a surprising observation that ran counter to expectations. I would like to see the second-generation data because I suspect the effect was transmitted through the females born during the famine. From embryology, we know that the egg cells of the next generation are laid down in fetal ovaries during the last trimester of pregnancy. I suspect they were imprinted with the same starvation message and thus passed it on to the next generation. To put it succinctly, the egg cell from which you originated was made while your mother was still in your grandmother's belly, and the environmental influences of that time will be imprinted and carried forward into your adult life.

In time, I expect that many more epigenetic effects will be documented; these will involve many aspects of physical and mental health. Even now, there are some known effects of prenatal stresses and toxic exposures. For example, alcohol consumption during pregnancy affects a child's learning abilities. I also wonder about the adverse epigenetic editing that must go on in the brains of babes still in the bellies of women who survive the stress of war zones. I fully suspect that the resultant post-traumatic stress syndromes will blight at least the generation that follows.

Genetics and epigenetics, like all scientific learning, are not abstract constructs emanating from effete ivory tower laboratories; it is important that we get the messages right. And to understand them, we rely on a century of basic work done with peas, fruit flies and laboratory mice. Once-abstract learnings have real consequences, and scientists must talk more urgently to communities and decision makers about those consequences. Today's strife seems to be announcing more pain for those who will inhabit the future. In short, it's never simply over when it's over.

A DIVERSION TO SEE YOU
THROUGH A PANDEMIC

You could lose your mind from considering the complexity of energy flows displayed in the classic "Chart of Intermediary Metabolism" that graces the walls of biochemistry laboratories. It's not a simple chart, but it is absolutely straightforward; just follow the arrows.

Our understanding of cellular metabolisms was gleaned from the study of mostly watery tissue extracts over the past 150 years. It's such a crude thing we do to tissues under study: we slice them, shred them, and treat them with solvents. Sometimes we force them along steep gravitational gradients in centrifuges and so much more. It's as though one had taken a bucket-loading tractor into a mall, scooped up a quantity of goods through a broken jewelry store window, crushed the items in a garbage compactor, and then attempted to reconstruct an electronic watch from the residue. It's nothing short of amazing that this approach to cells and tissues has been informative at all. But we have learned almost everything we know about how cells extract energy from their environment by using similar methods.

We teach biochemistry as though everything happens in conditions resembling what we extract from this watery tissue debris, but so much is different in the undisturbed cell. In the intact cell, there are multiple, highly organized and specialized, mostly two-dimensional molecular complexes that we call membranes, not unlike

67

those two-dimensional charts on the laboratory wall. Consider the inner membrane of a mitochondrion, where nutrition is turned into chemical energy. Here, we find the pathway that directs high energy electrons from sugar fragments toward their rendezvous with ATP synthase (to make the "gasoline of life"). According to Nick Lane, at University College London, the surface area of just one body's inner mitochondrial membranes may approximate to forty folded football fields; and that's just one kind of membrane, which produces more than your bodyweight of ATP every day. The ATP synthase enzyme that generates this energy is built into that same mitochondrial membrane, and the structure of the membrane channels the flow of electrons toward it.

Consider other specialized cell membranes. Some are fitted with pumps to preferentially excrete sodium ions while retaining potassium; others deliver glucose into the cell or redistribute hormones. The cytoplasm, far from being a bag of solutions, mostly contains more redundantly folded structures known as the endoplasmic reticulum and the Golgi apparatus; these have roles in the synthesis, processing, and transport of about 20,000 different proteins to their required locations. Whatever you imagine life to be, at an operational level, it is more like a production line moving along complex surfaces than an assortment of random molecules in a simple solution.

And that is not all there is to it. If we can accurately model the three-dimensional cell as a two-dimensional set of wadded-up membranes, why not further simplify the model? Why not make it a one-dimensional structure? That might be closer to the true state of other parts of our cellular interiors. That is what the nucleic acids— DNA and RNA—are. There are about two metres of DNA in each nucleus, the war room of the cell. That DNA holds our story, as if on tape, and the instruction sets for about 20,000 proteins, among other things. Messenger-RNA is the office memo sent by DNA from the nucleus (the head office); the mRNA to tell the microsomes (the factory floor), which are responsible for the synthesis of proteins, what to make next. These molecules are transporting digital information. We have made such rapid progress in understanding them because computers allow us to model these activities in detail.

Even after sixty years of wondering about it, my amazement at the jaw-droppingly wonderful arrangement we call "metabolism" persists. Yet, each step of this cascade can be stripped to its most simple thermodynamic form, which permits an accurate numerical description of the process. It comes down to the principle of Ockham's razor, which says that the simplest explanation—ie, the explanation that requires the fewest assumptions while still accounting for all the observable phenomena—is most likely to be right. In this frame of reference, the explanation can be amended without embarrassment if new or conflicting evidence comes to light. If the complexities are considered one at a time, they will all be found to have understandable, if not necessarily simple, solutions.

AN ODE TO DUST

It hadn't occurred to me until recently that dust might be a subject worthy of study. Then, I encountered a report concerning the state of the space suits that the American astronauts wore while on the moon fifty years ago. The report recounted that each astronaut had been issued two suits; one was checked out and then packed for the trip, while the other was repeatedly worn on training exercises in simulated lunar environments over several hundred of hours in the following two years. Recently, when planning began for a return to the moon, someone thought to examine the state of the original space suits. They observed that although the training suits were still in good condition, despite extensive use, those that had been used on the moon for about eight hours were badly worn and not suitable for reuse. The cause of the accelerated wear from the moonwalk proved to be lunar dust particles, which have very sharp edges akin to microscopic razor blades. In the absence of abrasion to blunt their edges, lunar dust had the rapidly deteriorated the integrity of the suits.

Where did lunar dust come from? The stuff we call "normal matter" constitutes only 4.9% of the mass/energy of the universe, and more than half of that is in the form of unstructured dust that permeates space inside and between stellar systems and galaxies. In short, dust occupies the universe. Earth sweeps up about one hundred tons of this incoming dust along its orbit every day. Most of the particles are very small, but there are a few larger ones; occasionally, even substantial meteors are encountered.

After the Big Bang, there was initially no dust. The first stars were formed purely of hydrogen and helium. Additional elements appeared from the early rounds of supernova explosions in the first one hundred million years and generated the first dust particles. Dust then contributed to the formation of subsequent generations of stellar systems and dust provided the materials required for the creation of planets. You can still see the leftover dark dust lanes of the Milky Way from your backyard without a telescope.

It was a major finding of the Kepler telescope that every star is likely to have one or more planets in orbit, and that the number of planets in our sky far exceeds the number of stars in the Milky Way. According to our current model of star formation, every birthing star is surrounded by a dense belt of rotating dust, where the particles, which are not drawn into the bodies of the parent stars themselves, become orbiting gravitational foci and some eventually form planets. This circumstellar dust cloud is of great interest because the products of previous supernova explosions include a large catalogue of molecules that are essential for life as we know it, as a bonus. Dust includes a Lego starter kit for life, including molecules of sugars, alcohols, amino acids, and fatty acids. Our own bodies are entirely made of the elements, which were themselves produced by the demise of stars that preceded us. Thus, it is plausible that something resembling life may be a recurring feature of the universe, and possibly even a standard feature in the histories of stars and planets.

As the universe matured, some places and certainly our planet, became friendly to life. We don't know how it began, but some of us are intrigued by the hypothesis that crystal surfaces provided the anvils that acted as the first enzymes that supported what eventually became life. The active centres of many of our enzymes contain metal ions: calcium, iron, copper, magnesium, and zinc, to name a few. Perhaps those ions are commemorating the past when they existed as active enzymatic centres on the surfaces of crystals. Studies of space dust point in that direction.

On the downside, dust makes observations of the most distant universe challenging. The air quality in observatory housekeeping must be maintained to minimize the problem of dust deposits on reflecting surfaces. As we seek to peer ever farther into the distance with ever-more powerful instruments, there is a cumulative amount of dust to be penetrated along the light column that may extend for ten billion light years or more. For purposes of viewing the universe, dust in space relates to objects of interest in much the same way that noise relates to signal. Considering how much progress we've made in sorting out signals, we are also overcoming some of the interference caused by dust. Since the launch of the JWST, we have been able to penetrate dust clouds using the infra red part of the spectrum and with spectacularly improved resolution of the images.

I'm fully expecting that advances in material sciences will suitably protect the next generation of astronauts from lethal attacks by lunar dust-bunnies.

A LITTLE FORCE MAY BE A GOOD THING

My first experience with science happened around the fifth grade, and it concerned magnets. Our one-room primary school on the prairies had few facilities, other than an occasional outburst of a creative teacher's imagination. On this occasion, she commissioned the students to bring various objects from home for an experiment. Mervyn volunteered that his father had a metal cutting file in his workshop, and he was given an envelope with instructions to bring iron filings from a rusty bolt he might find in the scrap heap behind his father's barn. Cecil's father had a substantial accumulation of discarded farm tractors in the pasture, so he was instructed to dismantle a magneto, a small hand-cranked dynamo used to start an engine before there were batteries, from one of them and retrieve a horseshoe magnet for the school. Back at school, teacher placed the magnetic poles under a sheet of paper, and we watched in wonder as the iron filings were sprinkled on the sheet and arranged themselves along the lines of force in the field. To this day, I struggle with the concept of "force." How does the magnet *know* there is a fridge out there? What is a force? What are these influences that cause matter to act on matter from a distance? Even though I am aware of the technical explanations of electromagnetism, the phenomenon still catches my attention.

It turns out that there are only four forces needed to explain the universe: electromagnetism, the strong nuclear force, the weak nuclear force, and gravity. We live every day with an awareness of the first and last of these, but we don't know nearly as much about the other two. The strong nuclear force prevents the positively charged protons in the atomic nucleus from repelling each other, and it is thus responsible for the stability of atomic nuclei and matter. The weak force is responsible for the fact that the Sun and stars shine.

How these forces originated in the Big Bang is a burning question among cosmologists. Do they have a common origin, or do they just happen to behave so cooperatively, despite separate beginnings that allow a coherent universe to exist? Some evidence, as well as some dominant theories, suggest a common origin. We use cyclotrons to accelerate nuclear fragments into high energy orbits and crash them together to create a facsimile of an early instant of creation. The theory that can be abstracted from these data suggest that following some sort of early primordial super force, the universe began to cool, and gravity separated from the other forces; henceforth, it was independent of the remaining Grand Unified Force which, upon further cooling, separated into the electroweak and strong nuclear forces. Finally, at lower temperatures approximating a relatively cool trillion degrees, the electroweak force separated to become the weak force and electromagnetism. All of that transpired in about a trillionth of one second. And these forces were then in place to guide the sculpting of the universe, allowing quarks to join forces to create the first atomic nuclei!

And here I am, still wondering how in Hell the magnet knows about the fridge.

A SLOW DANCE WITH DEATH

In our early life together, Lilianne and I bought a century-old farmhouse outside of London, Ontario, and progressively renovated it to accommodate our growing family. As I was ripping out the ceiling of the dining room one weekend, a fascicle of an old book, lacking title or evidence of authorship, dropped out from the floor above. It was a mid-nineteenth century medical book fragment that a child might have inserted through a crack between the floorboards of the bedroom above and it defined some medical problems of the day, with definitions of diseases and medical terminology arranged in alphabetical order. The fascicle that dropped on my head contained the C and D words.

In the nineteenth century, tuberculosis was known as "consumption." My attention was caught by the book's succinct definition: "A congenital disease of the working class."

When this book was published, the concept that infectious diseases were caused by bacteria had not yet been fully established, and people were at a loss to understand them. The term "congenital" pointed an accusing finger at the patients for having been born with a God-given weakness, for which they faced consequences that would dog them throughout life. With that understanding, there was not—and *could not be*—any social imperative to assist those so afflicted, except perhaps to provide comfort. Why exhaust oneself by assuming burdens that had no remedy, since the matter was out of human control?

It was only in 1884 that Robert Koch published his four postulates of the means of ascertaining bacterial causation of some maladies. In short, he held that the causative agent should be found in every case of that disease, that recovery from the illness should be associated with disappearance of the agent; that reintroduction of the agent should cause the disease to recur, and that the original agent should then be recoverable from the patient again. So, it was proven to be the case with tuberculosis. With that new understanding, it became possible to ponder possible therapies. However, the concept did not catch on evenly. In the first edition of William Osler's 1892 *Textbook of Medicine*, he was still uncomfortable with the idea that one form of tuberculosis, lymph node involvement of the neck (aka lupus vulgaris), was not congenital. In later editions, he accepted that all forms of the disease were infectious.

Tuberculosis was also known as the White Plague, due to the anemic pallor of many of its victims. In my childhood, my best friend's mother was taken to the Saskatoon Sanitorium with disease that had been diagnosed in a community screening exercise. She died there about fifteen years later, after she had lost faith in the ineffective nostrums of the day (in this case, pre-streptomycin) and ceased to cooperate with her medical caregivers.

Tuberculosis didn't behave like many other periodical or seasonal epidemics. It lingered for centuries among our forebears, like an *éminence grise* at the periphery of polite society, seeming to grin expectantly and whisper, "I'll get you, my pretty." Even a century ago, one of the two most common causes of death in the Western world was tuberculosis (the other being syphilis). We learned progressively that the incidence of the disease was strongly linked to poverty, nutritional status, and housing conditions—all factors that were strongly at play among the Indigenous population, as well.

When I came to London to complete my medical residency in 1970, my first clinical assignment was to the Sir Adam Beck Sanitarium at the western end of Oxford Street, where these disease forces were all still present. I was among the last resident physicians who worked there, as it was closed shortly afterward. By that time, the institutional approach to tuberculosis therapy was clearly obsolete, as we had several chemotherapeutic and surgical strategies for effective intervention, which allowed patients to freely circulate in society while under treatment. It was in that era that my ceiling opened and delivered its dated tablet.

Advances in technology provided many opportunities for us to recognize new options for personal and societal growth in all progressive and ethical dimensions. Without the lessons inherent in Robert Koch's tuberculous bacterial cultures and all the resulting knowledge about prevention and treatment, we might never have come to an awareness of our ethical duties to care for tuberculous patients or other social outcasts, including those with untreatable conditions. Those advances served to raise our sensitivity to situations where we can only hope for a cure. Why then, did we take so long to learn this lesson? Looking back, we can see that it was embedded in the teachings of Hippocrates 2,500 years ago, when he insisted that all diseases have natural causes, but we missed it.

Should we make the effort to ensure that drug addiction is the next disease about which we can feel hopeful?

THE MAGIC LAND OF HONEY

Beekeeping was my hobby for exactly thirty years. The choice was made for me when we moved to a farm, and the children declared their individual choices of pet. When they asked me for mine, I chose the one creature whose well-being would not be imperiled if I were kept late at the office. I knew nothing of what would be involved in beekeeping, as I had no mentors for that. Everything I eventually learned happened under the harsh tutelage of experience. Initially, I assembled a single hive from a kit in my garage and set it up behind the barn. Lilianne picked up a package containing three pounds of bees and a separate small cage with a queen and her retinue of worker attendants. There were instructions enclosed, and they said something like, "Open the container and pour the bees into the hive."

I was stung a few times in this initial venture—and more later on—but it was never so bad that I had to wear protective clothing. After about a hundred stings, I stopped reacting, except for a few seconds of transient light-headedness, almost like I was leaving my body. That feeling never went away, despite several thousand more stings over the years. One can learn to handle bees without stirring up the hive excessively, and most visits were uncomplicated. Over the three decades of our association, the bees taught me many things, of which a few come to mind now.

James and Carol Gould were entomologists in the seventies, and they shared a particular interest in insect intelligence. Bees captured their attention for many years. At that time, the American and Russian space programs were being launched, and public intellects were pondering the prospect that astronauts might encounter alien intelligences "out there." What were the odds that those aliens would be so different from us that we wouldn't be able to recognize them? What if they were aggressive?

Given this fear of the unknown, many people felt that we should search our own environments to find different models of intelligence that had not been considered heretofore. Insects, such as bees, ants, and termites living in colonies might offer some clues in that direction. A case in point was the then-recently discovered communications system that field bees use to direct hive bees to a newly discovered food source. They dance on the hive wall to indicate the distance and compass directions of the food, relative to the position of the Sun. Bees were also found to remember sites of previous food sources—a phenomenon that I have observed even several weeks after the initial exposure. There was also a report that bees can predict where a source that is moved day by day in a mathematically precise way will turn up next (eg, according to an exponential function), and they will eventually anticipate the location of the next food-drop. The question that may not yet be resolved is whether it is the individual bee or the totality of the hive that is the intelligent entity.

An amazing feature of bees is their social organization. There is a solitary, fertile female who rules the colony and lays all the eggs. Both the queen and the workers are females with identical genetics. When the workers feel that their queen is flagging in her egg-laying duties, they feed a concoction of royal jelly to a few larval bees. This concoction contains substances that bind to the larval DNA and reprogram these immature females to retain the ability to lay fertile eggs. We call the process by which the royal jelly controls the differentiation of female bees between queens and workers "epigenetics." I think of epigenetics as resembling a Post-it note inserted into a company procedure manual that says something like "Step 2 on page 23 is optional, pending a software revision." And this process is not limited to bees!

Indeed, in our personal embryology, each of us began life with a female body plan until day thirty-five, when an epigenetic phenomenon reprogrammed males to their destiny. The reprogramming is seemingly simple, and it involves the chemical addition of a single methyl group at a critical site on the DNA. But for that, this would appear to be a women's world. This particular methylation fails on occasion, and the event gives rise to chromosomal males with female external features. Think of it: but for that single methyl group in exactly the right place at exactly the right time, humanity might be structured along the lines of a beehive. Let us all give thanks for day thirty-five!

I wonder how such a restructured human race would differ from what we actually are. Would these ultimate drag-queens (men dressed up in women's bodies) be rounded up to form a slave caste? Would they be the solution to the problem of affordable childcare? How might the concept of 'human rights' evolve in such an asymmetrical society? What would war be like? These prospects, and much more, may be worthy of development in a serious sci-fi novel. And, if it feels transphobic, remember that's what the bees, ants and termites have done. Would/could humanity respond differently?

Each time I opened a hive, I learned something wonderful. The bees and I resonated together happily until age reduced my ability to lift those heavy, honey-filled hive boxes. That was more than twenty years ago, but I still sometimes miss those excellent teachers.

A WORD IN PRAISE OF WORDS

Ever since Charles Darwin deciphered the process of biological evolution, his work has been a challenge for those who wish to keep humanity in a separate box from all other biota, because the evidence points ever more strongly to our common origin with all life. Technical advances within the last decade alone now allow us to routinely sequence DNA from ancient bones up to nearly a million years old, and there is no generic difference in human DNA, recent or old, to distinguish us from any other mammal. There is one trait that may yet prove unique to humans, and that is our use of language. If only we could trace its biological origin!

There is a gene on chromosome 7, named FoxP2, that may provide a clue. This gene has been highly conserved in mammalian lines of descent for millions of years, and the human protein programmed by the gene differs by only two amino acids from that of chimpanzees. There is a human family in England with an inactivating mutation in this gene that has affected fifteen members over three generations; and all of them totally lack language ability while otherwise seeming to be within the normal range of intelligence. Imagine how difficult that assessment might be in the absence of any speech clues! Transgenic mice who have had their own version of FoxP2 DNA replaced by the normal human gene are reported to squeak differently from wild mice and learn mazes more quickly than their wild cage mates. The gene is being explored as a possible language-initiator and the enabler of fine control of the throat muscles involved in speech.

There is no doubt that animals, even those far removed from us on the evolutionary tree, use sound to communicate. One spring, when I didn't put the hummingbird feeders out on time, I was confronted by a bold bird who pointedly hovered in front of me to indicate that he remembered where the feeders had been last year, and then moved in front of my face to deliver a tweet that I heard as a demand to open the restaurant. What does this littlest of the dinosaurs' descendants and with a brain smaller than a grain of rice have in the way of speech capacity? Still, I understood the message.

I am trying to imagine the dilemma experienced by the first human who felt the urge to speak but totally lacked a vocabulary to do so. I think that this individual wanted to express a nuanced view that went into more detail than could be conveyed by growling, grinning, grooming, and showing teeth. Perhaps this early hominid merely needed urgently to swear to relieve a frustration! Isn't that how we, figuratively speaking, scratch our brain's amygdalae these days? From some such simple initial urgency, and with the accumulated experience of a hundred million years since we diverged from birds, we have achieved a measure of consensus concerning values and morality; we've learned to celebrate our lives together in songs, in inspirational books, and in blazing oratory. This much we know, and it is truly awesome. Nuances and subtle distinctions are basic to the creation of great literature, art, and the beauty of a tightly reasoned argument.

The path of progress is never smooth, and if you don't use the gift, you will lose it. Whatever has happened to considerations of beauty in our current public discourse, where lies and crass insults now dominate many conversations? I am trying to imagine the discussion at a forthcoming Research Ethics Committee meeting, when it will be reported that those same transgenic mice in the Animal Room bearing the human language gene have, after such a promising beginning, fallen into the deep insanity of alternate facts and convenient fabrications and become intent on intimidating each other with aggressive behaviours. There is a lot of growling, grunting, and showing of teeth. I hope that decisive action will then quickly be taken to eliminate those television news channels from their viewing options.

EPIGENETICS

Every school child knows that each one of the one hundred trillion cells that constitute a human body contains forty-six lengths of DNA that together are nearly two metres long—the largest polymers in the universe. The sequence consists of just four monomers, and the function of the intact molecules is to store digital information concerning the composition of the 20,000 odd proteins that sustain life. DNA is the cell's policy manual and instruction set, and is the custodian of its intellectual information. Mutations in the DNA (i.e., copying errors) may result in changes to protein composition, and they might compromise function in the progeny. However, there are many variations between individuals that occur in response to environmental pressures and are not related to mutations. I will provide three examples: the Agouti mouse, the learning disabilities in children whose mothers drank alcohol during pregnancy, and the multigenerational impact of starvation.

The yellow colour of the agouti mouse is a marker for a complex of features, including obesity and diabetes. The colour change occurs because the folding of melanin (i.e., the hair colouring protein) is altered by an environmental factor that binds to the gene. These changes are caused by stress to the pregnant mouse, and they are largely prevented by feeding methyl group donor vitamins, such as B12 or folic acid. Substances that stress the pregnant Agouti mouse will increase the frequency of this event. Bisphenol A, a normal constituent of plastic water bottles, is one such substance, and there may be others that have this effect. Acting out of an abundance of caution, regulators require plastic baby feeding bottles to be manufactured by processes that do not use bisphenol A.

The use of alcohol by pregnant women results in learning disabilities in their children. There are no mutations in the fetal alcohol syndrome, but the exposure stimulates the regulatory system to place a single methyl group on a cytosine base at a critical locus in the fetal DNA, such that certain information required to sustain normal functioning of the infant brain is blocked; this is rather like a Post-it note in a corporate policy manual that advises the reader to deviate from the official instructions for any reason. These Post-it notes can work both to the benefit of the individual, or they may be mischievous. At best, they are nature's way of coaching us toward a situation in our particular environment. But they may also compound problems.

I wonder what effects we will see on the next generation who were conceived and grown under stress—be it war, disease, or starvation. Might we be setting up for a future that will be primed by their mothers' prenatal stress, thus ensuring yet another war? Is that in some way how we keep on making similar mistakes as in the 3,000 years-long Israeli-Palestinian conflict?

Effects of this kind are subsumed under the title of epigenetics. Think of the environment as editor-in-chief of life's genomes and dictating critical overrides in the context of current external conditions. The epigenetic responses to stress possibly act to increase survival of individuals, albeit at a cost. Let us hope that something positive will come of our complex and tumultuous social situations.

COMMON SENSE

Around the turn of the century, we had a political regime in Ontario that sloganeered with, "The Commonsense Revolution." It turned out to be a period during which nothing much happened, except that some folks lost their jobs while the promised balanced budget never came to fruition. The term was emotionally loaded but never defined, even though it was used to convey a sense that truth was self-evident. And it's coming back again though nothing has changed. The problem with the term is that it cannot, by itself, evolve beyond the understanding of the day. The sense of "common" depends on yesterday's knowledge, and it does not take the flow of new evidence into account. The ways in which the term "common sense" is used in conversation implies that issues can be resolved merely by thinking about them in terms of the past and without testing for novelty. However, evidence often completely upsets the order of the day and brings us to amazing and novel new realizations. Here are three instances where common sense could not have been used to advance our understanding of the world. My list of examples is much longer than these three, but I am only writing an essay on this topic—not a book.

We've known for generations that bacteria are a major cause of disease, and that remains true. Common sense might say that we would be better off with fewer bacteria in or on our bodies, and that the current obsession with handwashing and antibiotics is entirely beneficial. Then we discovered some uncomfortable new facts. More than fifty percent of the cells that make up our bodies are bacteria; moreover, many of these are also essential to our remaining alive! More than a thousand bacterial species live in a healthy mouth. About 80,000 bacteria are exchanged in a one-second kiss. The demographics of our various warm and wet places are specific to each of us, and the microbial species that happily occupy some cavities would cause disease in another; staphylococci that would kill us in our blood stream live peacefully with us in our noses. The bacteria that are fermenting in our colons at this minute, producing tomorrow's bowel movement, are also making products that influence brain health and essentially tell us what we are allowed to think about! We are learning that children who grow up spending time in barnyards and getting dirty in parks are immunologically tougher and healthier than their tidier cousins who live exclusively in clean and supervised environments. Common sense couldn't have gotten us from the pre-1970s thinking to our present understanding; it took experimentation and the assimilation of a multitude of surprises to get us from "then" to "now."

By now, we've all put away the folding roadmaps that service stations once provided for free, and we put our trust in the GPS systems built into our cars. GPS is so convenient that we are not motivated to disbelieve its power, but do you know how it came to be invented? It wasn't by application of a common-sense extension of Newtonian physics, but through the unlikely mathematics of quantum mechanics and general relativity. In the sky, there is an invisible (to us) system of orbiting satellites. Our GPS detectors can simultaneously detect at least four of them and triangulate their signals to determine our position on Earth. You wouldn't have guessed from Newton's work that time passes more slowly on the satellites than it does on Earth, and the calculations that give us our precise location have to take that into account. Unless the GPS adjusts for the difference, the destination will not be found. Albert

Einstein first told us that this would be so in 1905, when he worked out the uncommon sense of relativity. 'Common sense' suggests that time is obviously invariant, but both relativity and quantum mechanics emphatically say otherwise.

Leaving aside these two examples as prologues, consider one more and, since I saved it for the last, the most anti-commonsensical of all. There are a few people in the world who essentially speak to each other in the language of mathematics. A conversation that began, more or less, with James Clerk Maxwell a century and a half ago continues today. He and Michael Faraday opened the conversation with contemplation of magnetism—a force acting on matter across a distance. Progressively, we learned that magnetism is not unique, but is one of the four forces that bind the universe together. Further, we learned through Albert Einstein that energy and matter are interchangeable in exactly the way that dollars can be exchanged for euros. Today we know that it is equally acceptable to speak of subatomic entities as particles or waves. During this search, we have found the equations that revealed "nothing" to be a "something." The same forces that attract the magnet to your fridge also exist in a vacuum and are statistically able to convert to matter; in short, they can create a universe from nothing. Common sense, where art thou?

Daniel Dennett has mused that the appeal to common sense is often no more than a failure of the imagination. No advances in our understanding of the universe have ever occurred without first breaking up the understood "common sense" of the day. Common sense may feel like a secure anchor from which to make "sense" of the world, and it seems to work sometimes over a narrow range of life's issues. But anchors that don't reach to the bottom, may only be weights that impede our progress in the exciting journey on which we are embarked. To apprehend the place in which we find ourselves, we must be always prepared for the surprises that await us around every corner.

ANOTHER IMPOSSIBLE THING

"...not for the good that it will do, but that nothing may
be left undone on the margins of the impossible."

- T.S. Elliot, Aunt Agatha in The Family Reunion.

In our lives, we experience only normal matter. But in the first instant of time, there were equal amounts of both matter and antimatter. Within seconds of the Big Bang, there was a massive reaction that destroyed all antimatter and all but one part in ten billion of normal matter. In that instant, almost the entire universe consisted of radiation. Why was anything left at all? That small, essential asymmetry of nature is still poorly understood and keeps some nuclear theorists awake at night. It might then come as a surprise to you that antimatter has come to play an important role in modern medical practice.

Nearly all the naturally occurring elements, the alphabet of matter, exist both in stable and unstable forms. The latter are known as *radioactive isotopes*. The isotopes of each element differ only in the number of neutrons located in the nucleus. Those isotopes with too many neutrons decay toward stability by emitting an electron and an antineutrino; those with too few neutrons decay by release of an antielectron (aka, a positron) and a neutrino. So, what use is that? How can we rationalize the taxpayers' cost of that research?

Most cancers have a greatly augmented appetite for glucose because they are not efficient at the complete oxidation of glucose to water and carbon dioxide. In order to live, they need to inefficiently ferment a lot of glucose to lactic acid. It also happens that these cells cannot distinguish between the natural form of glucose and a derivative that has been labelled with a radioactive isotope of fluorine (FDG). However, once inside the cell, the FDG is trapped in position and held there. A convenient radioactive form of FDG is labelled by a fluorine isotope with a ninety-minute half-life, and it emits a positron when it decays. We cannot measure positrons directly, but when one of them collides with a neighbouring electron, both are deleted from the universe. In their place, two gamma rays are emitted at 180 degrees of each other, and each with an energy of 511 kiloelectron volts. Detectors situated around the patient can recognize and record the coincidence and location of these two simultaneous rays. The accumulation of thousands of such events by the detector results in a three-dimensional image of FDG distribution and reveals the cancer. Further, we can use computer techniques to embed the FDG image onto an anatomically precise CT image, and this is known as a PET/CT scan. These are used to plan and monitor the details of, and responses to, therapy.

Since I left medicine, several new applications of PET/CT imaging have emerged, and there will be more, thanks to research by radiochemists. We now have agents that can help confirm Alzheimer's disease without an invasive brain biopsy. Brain PET imaging, in conjunction with functional Magnetic Resonance Imaging (PET/MRI), also has great promise to provide new insights into many aspects of brain function, including research into consciousness. There are also new agents for the imaging of ischemic heart disease that will lead to improved heart treatments.

All of this has become possible because we learned how to pull the Devil's tail (so to speak) and trick nature into allowing us to benefit from something that, in a sense, hasn't existed since the first instant of time. When I was retiring from practicing medicine, I noticed from my calendars that I had spent about a thousand hours writing reports and meeting with Ministry of Health staff to convince them of the efficacy of PET imaging. It took about another decade for others after me to convince them to fund these tests, and today's practices are at least ninety percent in accord with my recommendations from 1999. Jaded as I am about the bureaucracy, I am still thrilled by the science.

WHAT I SAW WHEN
I LOOKED AGAIN

Talk about collusion! A bit more than a billion years ago, two very different microorganisms plotted to take over and dominate the Earth in an ingenious way, and their regime exists to this day.

The primordial atmosphere of Earth was devoid of oxygen, since the high concentration of iron in the rocks rapidly absorbed all traces of it until plants outproduced oxygen sufficient to overwhelm the absorptive capacity of the planet. Oxygen was toxic to Earth's first life forms, and they evolved to use the energy of anaerobic metabolic pathways, such as sulfur dioxide in ocean-floor thermal vents. This stressed the early life forms and left them a limited number of choices: die, change, and/or make a compromise that would permit survival under these new conditions. One possible solution involved cooperation of oxygen-intolerant species with those that had evolved to handle this new toxic gas. Cooperation can only work if each party brings something to the table. In this case, the Archaean microbial partner provided the management skills to 'run the office' and the cell nucleus, while the Bacterium had the secret for living with oxygen in its mitochondria. Presto! Behold the modern eukaryotic cell that ultimately made multicellular life (including humans) possible!

How do we know that this is even plausible, let alone true? We can learn a lot from examining the intruder who has taken up residence with us. First, we note that mitochondria have their own

DNA (mDNA), indicating that they were once free-living organisms in their own right. To this day, they retain a few active genes that code for critical aspects of their role in us. Secondly, mDNA is different from our own chromosomes in that it is of bacterial type; being circular, rather than a linear, molecule. Thirdly, the first codon that signals the beginning for the translation of a mitochondrial gene into a functioning protein is a modified amino acid—formyl-methionine, as in the case of bacteria—whereas chromosomal genes use methionine for this purpose. Finally, mitochondria are not limited to replicating only with cell division, and they use bacterial fission to increase their numbers, quite independent of what the cell nucleus may be doing.

It has been demonstrated by studies with labelled DNA that over time, many mitochondrial genes have migrated into the cell nucleus and joined themselves to chromosomes. This has left mitochondria without the ability to live any longer as independent bacterial species. In the human case, there are only twenty-one active genes still coded in the mitochondrion, compared to about 23,000 active genes in the nucleus.

What are the Archaea, the invaded lot with elegant nuclear chromosomes, getting out of this relationship? The answer is "life." Every glucose molecule (six carbon atoms) that enters an anaerobic cell is rapidly broken down to a couple of two carbon fragments with the net production of four molecules of the cell's high-octane fuel, ATP. The task that the mitochondrion has assumed is to take these leftover acetyl groups and make them react with oxygen to produce water and carbon dioxide, along with thirty-four additional ATP molecules. The partners have become totally dependent on this cooperation. It has made possible multicellular life as we know it today.

There are some interesting considerations that drop out of this. For instance, mDNA is inherited differently from the chromosomes in that fathers play no role in its transmission; it's all from mothers to children. Since there is no need to randomly recombine them with the paternal mitochondria, the mitochondrial chromosomes are inherited intact, and it is possible to trace them back to near the beginning of our species, where we have found that "Mitochondrial Eve" was in Africa more than 200,000 years ago.

As a practical matter, there are sometimes mutations in mDNA. There is some protection against this occurring because there are many mitochondria in each cell, and the mutants tend to be diluted and displaced, but when they occur and are transmitted, the result can be debilitating, or even lethal. The offspring may have problems relating to the production of energy (ie, myopathies). This has recently been overcome by the model of the child with three parents: one mother who provided the defective egg, another woman who provided the healthy mitochondria, and the father. Together, and in a Petri dish, this trio overcame the effects of the mutation to produce a healthy child.

There are critics who rale against the materialism of science, but I suggest they be silent and allow themselves to listen to the grand symphony that is Nature. For three hundred years, we have put up with folks like John Keats, who complained that Newton, by reflecting light through a prism, had "unravelled the rainbow" and seemingly reduced its wonder to a pile of multicoloured threads. I think that at one period in history we gazed on what we did not understand through spectacles that paralyzed our understanding. We were looking but not truly seeing. In order to truly wonder, we must reach beyond the refuge of intellectual paralysis. New insights can be terrifying on account of the new possibilities and moral responsibilities they raise. There is much new territory to explore. I am trusting that society will be up to the challenges.

BELIEVE IT OR NOT

Microbial life first appeared on earth about four billion years ago—ie, within five to six hundred million years after formation of the planet. However, multicellular life did not appear until about one billion years ago, and the unicellular life forms had the place to themselves for the greater part of Earth's history. What were they doing all that time? It is becoming clear now that, far from sitting on their hands, so to speak, while waiting for us to arrive, they were busy infiltrating every possible biological niche and adapting themselves to fit to a wide variety of conditions. The process of adaptation caused them to evolve into the incredibly large array of bacteria, archaea, fungi, and algae that we find thriving in some very surprising and extreme environments today. I doubt that we have, as yet, catalogued them all, but here are a few that seem to me to be amazing with respect to the growth conditions they adapted to.

I was introduced to the *Deinococcus radiodurans* fifty years ago in a productive time for radiation biology. This accidentally discovered bacterium is noted for its ability to survive large acute doses of ionizing radiation. Following a dose of 3.5 Gray (Gy) of X-rays, about a thousand times more than a human could survive, it stops multiplying for about three hours, during which time it repairs up to a third of its DNA and then resumes growth, as though nothing had happened. On one occasion, I had to amend the conditions of its growth from the liquid medium to the solid surface of an agar plate to make a different measurement. To my astonishment, there was no cell death at doses below 12 Gy. Some scientists then asked whether an organism such as this could possibly have been the vehicle by which life from another planet might have been seeded onto the Earth from outer space. The question remains open to the present. The *Deinococcus* was later shown to be very facile at DNA repair through a mechanism we now know as recombinant repair. It uses the information sequences of additional identical chromosomes as a template to repair damaged regions.

Acidothiobacillus thiooxidans is a bacterium that requires oxygen to convert sulfur, its energy source, to sulfuric acid. It prefers a pH of 2.8, which is similar to car battery acid. It also may be found in sewer pipes, where it creates an acidic effluent that contributes to the progressive dissolution of concrete. In caves with atmospheric hydrogen sulfide, it grows on the walls and ceiling to form a mucous-like film known as snottites. Where does one find battery acid in nature? Your best bet would be in volcanic crater pools and volcanic caves, where subterranean sulfur dioxide gas is seeping upward and dissolving in the water. Today, there are fewer simmering volcanoes than existed in the remote past; it seems likely that back then, the Bacteria were carried along with the dust of eruptions across the Earth, sometimes landing in another appropriately acidic pool.

There are about ten thousand active volcanoes on the floor of our oceans, mostly along the Atlantic rift and the ring of fire surrounding much of the Pacific Ocean. They are often at several kilometres depth. When these volcanoes erupt, the water temperature in their vicinity rises well above 100°C (i.e., the boiling point at sea level) but the water pressure prevents active boiling at depth. Thermophilic life forms, including species of crabs and tubeworms, lie dormant on the ocean floor, but they become active as the temperature rises. They grow and multiply at water temperatures of up to 122°C. Other species of thermophilic bacteria have been found in volcanic sedimentary rocks in Antarctica, of all places. The enzymes of thermophilic life are active at high temperatures, and DNA polymerase from this source is used in laboratory DNA sequencing to speed up the process (the polymerase chain reaction, or PCR). Thermophilic bacteria are currently regarded among the oldest life forms on earth, and the study of their metabolism can inform us about the nature of early life.

One of our great surprises has been to learn that even solid rocks are permeated by life. We call these organisms petrophiles, and they have been found in the deepest boreholes that have been made so far. They live by metabolizing micronutrients in rocks and crude oil, where it is available. Japanese researchers reported recovery of thriving bacteria from a coal seam located 3.5 kilometres below sea level, under the floor of the offshore Pacific Ocean, where they were reducing methane to generate energy.

In years to come, we will eventually learn about the conditions that exist on some exoplanets. We will also learn whether something resembling life as we know it might be possible there. The basic building blocks of life have been found in space dust, on meteorites and the like, and it seems possible that similar biochemistry to Earth evolved elsewhere. I hope we will be alert enough to recognize it when we see it. Stay tuned to this channel.

THE BULLET-PROOF SHOELACES

I was complaining to a shoe store clerk about the inconvenience caused by an unscheduled lace break. She offered a pair of braided Kevlar laces at an increased price. That's how I bought the bullet-proof laces for my work boots. "These are guaranteed for life," she said. "Just keep the receipt."

"Whose life would that be?" I asked. "The laces', the work boots', or my own life?" She didn't get it.

I can now report that these laces have, so far, outlived the original boots, as well as two succeeding pairs. They are now installed in third-generation boots and have still shown no overt signs of wear. They are showing signs of outlasting even me. Now I'm beginning to think that I ought to have bargained for a more durable guarantee so they could one day be featured as valuables in my estate. I am imagining my heirs fighting each other for possession of my near-eternal laces; I can even see the ultimate possessor's chagrin when they eventually do wear the laces out and discover that I lost the non-descript cash register receipt required for a free replacement.

I'm not sure where this is taking me, except that I'm reminded of a popular professor who delivers frequent public lectures. This professor sometimes appears at the podium dressed in formal evening wear, except for tennis shoes, which have been infiltrated by brightly fluorescent yellow laces. To this, I respond, "I WANT THOSE!" But alas, I have squandered my budget on durability and have nothing left over for a mere fashion statement. As is true elsewhere in life, I have learned that objectives, no matter how brilliant the concept or how heartfelt the quest, are quickly absorbed into the fabric of the mundane once realized. The children ask with puzzled expressions, "What's the big deal?" The facet of myself that I have discovered in this discourse is that it is possible to be simultaneously both safe and sorry; I should have framed that damned receipt.

Sic transit gloria mundi.

ANTICIPATING THE INEVITABLE

It is impossible to predict a moment when one might fall in love. And so, once again, I find myself surprised to discover that Orion has become my favorite stellar constellation. I think almost everyone knows Orion in our southern sky. It has three stars in the belt, four more marking shoulders and knees, and a diffuse blush near his nether regions, supposedly representing the sheath of his sword. The latter is actually a nebula—a star nursery where many new stars are being born and finding their futures among the debris left after many previous supernovae. However, my attention has recently been focused more on Orion's, possibly arthritic, right shoulder. The star he wears there is Betelgeuse, and it is distinguishable to the unaided eye by its redness.

Betelgeuse is in difficulty, as we know from observations over the past two hundred years, since the astronomer and composer William Herschel noted that there were times when it was the brightest star in the sky and other times when it was not. Betelgeuse is in the red giant phase of its life—ie, it is hundreds of times larger than the Sun and has a circumference that approximates the orbit of Jupiter. Betelgeuse is running out of fuel, and the struggle between the thermal energy that has kept the star puffed up and the gravitational energy that wants to collapse it into a white dwarf, neutron star, or black hole is coming to a crisis. Betelgeuse has run out of hydrogen, and it has switched to fusing helium into carbon. That progression will end with explosive effect on a scale we have not seen in our galaxy since the time of Johannes Kepler.

It will happen one night—perhaps tonight, or maybe not for a thousand years yet—that Betelgeuse will brighten up to become visible to the naked eye even during daytime hours for a period of several months, and the night sky will be about a hundred times brighter than the full moon before it gradually fades to invisibility over a couple of years.

There have been about six such stellar explosions in our galaxy in recorded time, the nearest of which (the Crab Nebula on July 4, 1054) was more than ten times farther away. The Betelgeuse supernova will be the nearest supernova ever. The visible light energy released from the star, spectacular as it will be, will constitute only about one ten-thousandth of its total energy release. Distance and our atmosphere will act to protect us from the intense ionizing radiations. Further, the radiations will not be emitted equally in all directions because all stars rotate and as they collapse; they spin up to a very high rate to conserve momentum, the way spinning figure skaters do when they pull in their arms. While the skater spins at no more than about five times per second, the star will attain rates of thousands of rotations per second, causing the radiation to be mostly emitted in two narrow beams along its axis of rotation. The direction of Betelgeuse's axis is about twenty degrees off from our line of sight; so, it seems we will dodge this beam, should it happen while we are still here.

Can we predict when this supernova will occur? In a word, no. In one sense, it may have already occurred. The distance between Betelgeuse and Earth is 640 light-years. If the star actually exploded when Columbus discovered the Americas, then we wouldn't see anything for yet another half century. On the other hand, the explosion may not occur for a few thousand years yet. But, whenever it happens, we should first detect it in our neutrino laboratories (one of which exists at the bottom of the nickel mine in Sudbury) and the gravity wave detectors that are now being deployed around the world. Within hours after the explosion, the visible light emitted as the star physically disintegrates will reach Earth. If we are here to see it, Betelgeuse's death will be the fireworks display of all time.

Every evening, when I take my dog out for our final patrol of the garden perimeter before bed, I stop to look for Betelgeuse, just to be sure it is still there. As grand as the events culminating in its death will be, I anticipate that I will miss it when it goes.

CANCER

Who among us has not hurt ourselves, causing a bit of bloodshed—enough to warrant a timeout and a band-aid (and perhaps a tetanus shot, too)?

When this happens to me, I usually contemplate the mess I've just made and feel a sense of nearly blinding awe. Consider what has been set in motion. The injury sparked an outburst of molecular messengers from the damaged tissues that, in turn, launched a burst of reparative activity. This resulted in the laying down of new fibroblasts, the construction of new blood vessels, and the covering of the wound with new skin cells. And then, right on cue, when the job was done, it all stopped, the scab fell off in the shower and the incident soon forgotten. Talk about tight control! What could have happened if an element of that response system had failed to fully respond on cue? One possibility is cancer.

In large terms, there are three drivers that operate in the evolution of a cancer. Think of them as initiators, growth factors and disseminators.

The initiators are also known as oncogenes. These are the genes that were once involved in the normal repair response, but that have undergone random mutations and have lost their fine controls. These all derive from dominant genes, and carcinogenic change is the result of a single mutation. Every adult has millions of cells with oncogenes that are revved up and ready to get on with the business of cancer. So, why are we still here? The short answer is that oncogenes are not the whole story.

The oncogene's impulses are subject to control by an array of tumour-suppressor genes. These invariably are recessive genes that require two independent events in order to be completely inactivated. One example of suppressor activity is illustrated by bowel cancer. The initial manifestation of this cancer is some minimal thickening of a patch of mucosal cells on the gut wall that the pathologists would characterize as metaplastic—ie, not quite a cancer. Eventually, a suppressor gene in the patch is inactivated by a mutation, and at the next examination, there may be a polyp with a small cancer (*in situ*). The process repeats randomly, and at a later stage, there will be a local bowel cancer. Finally, when all the brakes are off, there will be a progressive cancer, complete with invasion of lymph nodes into neighbouring tissues. A similar story can be told of breast and ovarian cancers, where there are a couple of suppressor genes (BRCa1 and 2) in play. Since the mutations occur at random, the process may take a couple of years to play out to a full-fledged cancer lump, and that is why early diagnosis plays such an important role in successful treatment.

The story so far has explained the initiation and local growth of cancers, but not how cancers may spread around the body. For that, we need to introduce a third concept; that of cancer stem cells. Let's drop cancer for a moment and focus on stem cells in general. Embryonal stem cells have been in the news for a couple of decades, sometimes controversially, on account of their role in conception. Critical to this discussion is the understanding that all tissues of the body were nascent in that first cell that was generated upon fertilization. Over time, as the embryo developed and became multicellular, the potential of individual cells became progressively more limited, and certain regions of the genome were locked out of further participation to ensure that skin remains skin and never becomes liver, or retina, etc. However, differentiated mature cells of any organ (eg. skin) retain the inactivated genetic information that describes the other organs (eg, retina). In order for some embryonal cells to end up in the right places in the baby's body, they need to physically migrate from their place of origin (eg, melanocytes from central nervous system to skin); upon arrival on site, that controlling genetic sequence was turned off and chemically marked as "*Danger! Do not open under any circumstances!*"

You can imagine the cancer cell, which has already escaped some constraints of growth, looking through its genetic closet, finding this Pandora's box containing the secret of migration throughout the body. Something like this is what triggers cancer to spread. Most cells in a cancer lump will never divide or go anywhere again; cancer stem cells are rare—perhaps one in tens of thousands of cells develop the capacity to travel.

It is noteworthy that while oncogenes and the progressive failure of tumour-suppressor genes happen as the result of mutations (i.e., changes in the DNA), cancer stem cells arise and operate in response to epigenetic forces. That is to say, their behaviour is moderated by environmental factors that operate from outside the control of DNA. That, in turn, means that they may require different therapies for control. Cancer stem cells are, in fact, resistant to treatment by many mainstream therapies, which are directed toward shrinking the tumour. Much work is in progress to identify new therapies for cancer stem cells. From surveys of thousands of potential chemicals, Salinomycin was identified in the laboratory, and said to be a hundred times more effective than leading anti-tumour chemotherapies. Ironically, this compound is widely used in agriculture to control coccidiosis in chickens, and there is a laboratory joke that speculates about the therapeutic effectiveness of KFC or Swiss Chalet cuisine.

Will we ever definitively solve the cancer problem? Some hopeful observations come out of studies of elephants and naked mole rats.

First, a mature bull elephant has a life expectancy of eighty to eighty-five years and weighs in at 13,000 pounds (which is equivalent to about eighty humans). If its risk of cancer was equivalent to ours on a cell-by-cell basis, elephants wouldn't exist. But cancer is extremely rare in elephants. We now know that they have an additional DNA repair mechanism that removes mutations before they become fixed in the genome. We have the same gene as well, but it is inactive in us on account of another mutation that has broken that switching mechanism. To turn it on, we would need to reverse-engineer the switch; in principle, we could do this. However, we have learned to be wary and aware that the law of unintended consequences may result in more harm than good. Will we ever do it? Perhaps, but not likely in this lifetime.

Second, the naked mole rat is just that: a pink, hairless rat that lives in underground colonies in Africa. It is about the size of a guinea pig, and animals of this size generally have a life expectancy of four to six years. But some naked mole rates have been kept in captivity for more than 30 years without showing a tendency to either age or develop cancer. The females remain fertile throughout their lives, which can last up to 400 years in human terms. It is known that they have some unique features in their hyaluronic acid, which is a component of the adhesive molecules that connect cells into functioning tissues; these features mitigate against tumour growth. It would be a nice bonus to solve aging along with the cancer problem!

The message we should take from this is that any cancer that is prevented is associated with one hundred percent survival. Those that occur despite our best efforts are treatable with a progressively improving outlook. The future holds out possibilities that we don't yet know how to grasp, so never give up hope.

DOSE-RESPONSE CURVES

Judging by the weight of the name they gave him; one might imagine that his parents were banking on him to anchor (or sink!) the family ship. At the baptismal font in 1493, he was named Philippus Aureolus Theophrastus Bombastus von Hohenheim. He was educated in medicine by his physician father and had a peripatetic life while he simultaneously practiced medicine and irritated the medical authorities of the day, who often interfered in his practice. His personality continually got in the way of his innate brilliance, and it is thanks to him that we have the adjective "bombastic" in the English language. He changed his name to Paracelsus (possibly as contrast to Celsus, a second-century philosopher) and he is better known in history by that name.

He became known for his insistence on the importance of evidence, a penchant that infuriated the medical power structures which were still insisting on the *prima facie* authority of Galen and Vesalius to form the basis for all medical learning. Paracelsus was the first to insist on exact dosages of prescribed medicines in therapeutics. He became a prolific writer but was mostly blocked from publication by the medical establishment. Ironically, he died mysteriously before the age of fifty, possibly of poisoning, perhaps at the hands of a cabal of jealous physicians who were concerned that his teachings were interfering with their business model. (It was perhaps an early attempt at a medical disciplinary committee action, but driven purely by emotions and without evidence?) His aphorism, by which we most often remember him today, is: "Everything in the universe is toxic at the right dose."

The dose-response curve is a fundamental tool of modern toxicology, and Paracelsus would have loved it. A dose-response curve may have different shapes and require different interpretations. A linear curve indicates that a substance is similarly effective or toxic over a wide dose range. For example, cigarette smoking is toxic in direct proportion to the amount consumed; for such products, the most rational public health response should be to outright ban the product (in this case, tobacco). Other examples of possibly linear toxicity might include mercury or lead, which are associated with brain injuries.

A flat linear effect curve is seen on the beaches of Kerala, India, where thorium in the sand gives radiation dose measurements up to about fifty times that of the North American plains. However, studies of local health status show no effect on cancer incidence. This is so because the radiation is short-range beta emitting and is largely absorbed by the soles of thick-skinned feet or shoe soles—even flip flops.

There are also S- or J-shaped curves for other agents; some materials may be not toxic, or can even be essential for life, within certain dose ranges. In these cases, we seek to define thresholds of injury. Take the case of oxygen's S-shaped dose response. I was a paediatrics intern sixty years ago, before the advent of intubation and ventilation technologies for newborn, premature babies. At that time, the standard treatment of infants with immature and poorly inflated lungs was to place them in incubators, flood the enclosure with one hundred percent oxygen, and hope for the best. When doing so, we also informed the parents that this possibly-lifesaving, treatment was also associated with a risk of retinal fibrosis and blindness. The choice was between a blind child or a dead one. With modern techniques, the expansion of the lungs and oxygenation can be addressed somewhat independently, and toxic oxygen doses are no longer administered.

In 1948, David Leakey was a young graduate student at an American midwestern agricultural college, where his supervisor encouraged him to study a new class of chemicals that we now know as herbicides. For his thesis project, Leakey set up a series of standard weed plots and treated them all with progressively more diluted concentrations of the first post-war herbicide. At the company's initially recommended concentration, he got a one hundred percent weed-kill; at lower doses, it dropped progressively and eventually became undetectable. It is a testimonial to Leakey's observational skills that he then observed that there was a still lower dose range at which the herbicide actually became a growth stimulant—an effect he named hormesis. This is an example of a U-shaped response function. Over the remainder of his long career, Leakey showed that the concept of hormesis is broadly applicable to many classes of toxins, including ionizing radiations.

Misapplication of dose-effect curves of different shapes has had wide-ranging impacts on the public perception of risk. In 1946, when Herman Muller received the Nobel Prize in Medicine for his work on the mutagenic effects of ionizing radiations, he proposed a linear-no threshold (LNT) approach to radiation safety based on his earlier work concerning the incidence of eye color mutations in irradiated fruit flies. He ignored the fact that his graduate student at the time of the award was repeating that same work with a more sensitive methodology, and was recording a threshold effect.

Muller was well intentioned; of that, I have no doubt. All scientists at the time shared a sense of guilt over the bombings of Hiroshima and Nagasaki, and they were actively looking for socially redeeming ways of using nuclear science. Despite having endangered his graduate student's career, Muller seems never to have modified his position, and his elitist, authoritarian stance on the subject limited the freedom of subsequent standard-setting bodies to make recommendations based on other insights. LNT caught on with the worried public and was even perversely used by anti-nuclear groups in their campaigns, both rightly and very wrongly, to validate a sometimes-frenzied hostility to the ongoing development of nuclear sciences. Even though we now have mountains of data about radiation dose-response curves that clearly show threshold, and even beneficial effects, we still have anxious people, such as folks on the west coast of Oregon in 2011, who worried about possible radiation exposures to their radishes across the Pacific Ocean from Fukushima. At that time, some worried people took prophylactic iodine against expert advice, and they actually increased their radiation dose because iodine is provided as a potassium salt, and potassium comes with an irreducible taint of radioactivity that persists from the Big Bang of creation. The problem ultimately factors down to the management of risk perception, as well as of the risk itself.

Paracelsus would have demanded a pause in the debate to define terms, something that few in our recent generations of activists have bothered to do.

DUPLICATION AND INFORMATION

My midnight brain recently went roving and snagged in the dark onto a fragment of Umberto Eco's essay concerning the utter futility of maps drawn to full scale. It is intriguing that a freehand sketch— the sort that was badly drawn on a coffee-stained napkin—could be adequate to guide me through town to your house. However, if the map is enlarged, and more information added, one immediately encounters the problem of how to fold it without losing sight of the destination. The ultimate enlargement—that is, on the scale of the city itself down to the finest detail—would amount to its duplication and would totally negate the purpose of a map. How odd it is that on most occasions, less information, pruned to address necessity, is more useful than the whole of reality.

In theoretical physics, explanations at the extremities of the universe—eg, the origin of the Big Bang and the ultimate collapse of matter into black holes, terminate in singularities. These are perhaps best thought of as the classical "divide by zero" error, where all values run off to infinity and terminate the math—but this only at the beginning and the end. How odd that maps at full scale have a feature that also bears at least a metaphorical resemblance to a singularity, where the maximization of information only frustrates the point. Smaller-scale maps may be useful, but the reality-sized version is not informative; it is only a duplication of reality.

It occurs to me that Eco was playing a game with our minds when he took us (or, at least, me) down this rabbit hole. After all, he was a semiologist by profession, which is, at least for me, a bewildering field of abstractions concerning the links between things and their names. However, his magical skills with images and their names are well displayed in his complex, profound, and often witty novels, such as *The Island of the Day Before.*

In a final attempt to extend Eco's argument, possibly beyond his intent, I will suggest that his semiotic game has revealed that the duplication of objects is distinct from developing information from them. For centuries, scholars laboured in monasteries to reproduce copies of rare books and illuminated Bibles; none of their efforts availed them one bit, and society's state of enlightenment was not increased over many lifetimes. In short, throughout those centuries, they were duplicating what they already had without mapping toward any new territory. Memorization never produced scholars or poets—only mimes. There are times when the Coles Notes may be more informative than an accurate and complete Shakespearian folio.

If I learned anything before sleep overtook me again, it was that the most intriguing part of a map is found at its edges, because there the imagination is at liberty to propose new territory. Thank you for that, Umberto.

DILEMMAS

I believe that everyone is familiar with the textbook ethical conundrum in which a runaway trolley is about to run down five unsuspecting track workers. You have the option to throw a switch that will divert the trolley to another track, where only one worker will be killed. Would you throw the switch and take responsibility for that one death?

In the movie *The Imitation Game*, which features Alan Turing's work, there is a scene that occurs following his team's realization that, after four years of hard work, they have broken the German military codes. Immediately thereafter, a German U-boat message that it is about to sink a named British supply vessel is intercepted. One of Turing's team declares that his brother is aboard that ship, and he advocates for immediate action to break up the pending attack. However, his superiors declare that the Germans would then immediately know that their codes had been deciphered, and this intervention would void all the work required to break the codes and frustrate their war effort thus far. In the movie, the brother was lost. So far, no one has suggested that Turing's team acted other than honourably on that occasion.

Let's recreate this moral dilemma in a hospital. As we make rounds, we notice that the two patients in the first room are suffering from renal failure, and each needs a kidney transplant. The patient in the next room needs a liver transplant, another needs new lungs, and the last needs a heart. But wait a moment… Didn't we notice the lone healthy-appearing visitor in the waiting room when we came in? Are five lives saved at the cost of one a good trade? Dare we even think of snatching the opportunity? It is clear at once that one answer does not serve all supposedly parallel scenarios equally well.

In the late 1980s, when I chaired our hospital's medical advisory committee, the local newspaper published an event that got the community's attention. On a certain night, the airport turned on its runway lights at 4:00 a.m.; shortly thereafter, a small aircraft landed. A man carrying a box was observed exiting quickly from the plane and entering directly into a waiting limo on the tarmac. He was then driven with a police escort to Victoria Hospital. The man entered the hospital with his box and re-emerged in some haste about two hours later; he returned in reverse sequence to his waiting aircraft, which took off immediately. The journalist asked: "Who was this man, what was his business, and where had he come from? What was the box about?" The fact that this all transpired at night made the entire episode even more mysterious. As it turned out, these were the earliest days of organ transplantation, and the mystery man was merely a surgeon from Pittsburgh who had come in the night to collect an urgently needed liver donation.

It didn't help matters that Robin Cook's novel, *Coma*, was popular at the time, and the public were widely aware of its plotline. In that story, a physician at a Boston hospital is developing organ transplant therapy and resorts to switching anesthetic gases during routine surgery to preselect the most desirable organ donors for his program. Thus, we had at least one fictional physician who was willing to take out an unsuspecting "donor" for the sake of others. That episode, and the public's awareness of *Coma*, caused us to review, tighten, and formalize our protocols around organ donation procedures.

It's not a big stretch of the imagination to think that other things might go off the trolley track in transplant medicine; there are organs for sale in some places, and a desperate patient who goes to such lengths for a new organ may choose not to look too deeply into its provenance. Impoverished individuals have been known to sell a kidney without knowing what was involved. We are also aware there is organ retrieval in the aftermath of executions in some countries. Does the market for organs then influence the incidence of executions?

Back to the trolley: what if the trolley is, in fact, a driverless car, and the action to be taken depends on the ethical awareness of the programming team that designed it? In this situation, I think we have a few new options. There will be times when the speed-of-light reactions of an AI-driven device will serve to alter the outcome. Once upon a time, in the name of safety, we sent a flagman on foot ahead of the motor car to warn the oncoming horse-drawn coaches. We can't just proscribe AI-driven transport in our time; instead, we will have to deal with the ever-renewing challenges of ethical dilemmas. The answers won't always be the same.

FROM NUTRIENTS TO MAGNETIC FIELDS

The most common nutritional deficiency across the plant and animal kingdoms is nitrogen. The deficiency manifests itself as the yellowy shade of an unfertilized wheat field, as well as in medicine, where it is known as kwashiorkor, or red boy disease, in starving populations I saw a case involving a child from Gaza on a recent TV newscast. This may seem surprising, given that seventy-eight percent of the atmosphere is nitrogen. Why didn't evolution develop pathways for cells to directly capture and entrain this essential element into our metabolic pathways?

In fact, nature has provided two very different solutions by which to deliver nitrogen to biology's door. The first is via the Rhizobium bacteria, who live on the roots of legumes and are equipped with a chemical pathway to produce nitric acid, the mechanism that is responsible for about seventy-five percent of nitrate production. In this form, nitrogen becomes bioavailable, and it can be taken up by cells and used to produce essential molecules, such as amino acids and nucleic acid bases.

The second mechanism, amazingly, is via lightning strikes. When the electric charge in a lightning bolt traverses the atmosphere, it transiently heats a column of air to more than 3,000C and forms a volume of nitrogen free radicals that will interact in milliseconds with almost anything they encounter, including oxygen atoms, to make nitric acid. These newly formed nitrates then dissolve in falling rain and contribute to fertilizing the soil. Altogether, these two mechanisms account for an annual worldwide nitrate deposition of about a kilogram per hectare. Until the human population grew to its present proportions, this was enough to maintain the level of plant and animal health that a smaller population required. In recent times, we have had to add chemically produced nitrate fertilizers in order to achieve the required production of grains to the market.

It turns out that there is more to this story of lightning than just a free lunch. The energy discharged in the bolt can reach to the range of the discharges of a large experimental nuclear accelerator. In the process, electrons are indiscriminately torn from atoms encountered in the air column and transient production of molecular and atomic fragments including free radicals as well as positrons and free electrons occurs. The positrons (electrons with a positive charge) have a short half life and are neutralized before they have diffused more than a few millimetres from their point of formation, but the electrons are directed by the earth's magnetic field to accelerate at relativistic velocities along the magnetic lines of force to the poles. This phenomenon first came to light when an orbiting satellite happened to be directly in the path of such a discharge and was briefly taken out of commission by the damage. Along the way, some of these energetic electrons will also collide with additional nitrogen atoms, thereby producing a few more nitrates as they go.

Let's also take a moment to applaud thunder. The bang comes about because the heated cylinder of air expands faster than the speed of sound; in short, thunder is the primal sonic boom of the planet. For millennia, our ancestors interpreted thunder as an indicator of God's (or the gods') displeasure with our behaviour. ("You guys down there! Behave yourselves, or I will make it rain so hard that the Earth will be flushed clean!") In Carl Sandberg's account of "The War Years", Abraham Lincoln spoke critically of his political opponent, who was known as a canny merchant, for having mounted lightning rods on his house "to protect his guilty conscience from an angry God." Today a thunder growl just means that one should leave the golf course.

Worldwide, about two hundred thunderstorms daily bombard the atmosphere over our heads, each salvo providing a distributed benefit to both cultivated and natural habitats. The wonderment is breathtaking, and we are no longer so worried about the moods of the gods. I've come to think of thunder as the Earth applauding yet another successful delivery of lunch to the delicate, yet enduring, marvel of life.

I am constantly amazed that it takes so many different threads to come together to create the beautiful tapestry that is the universe.

FROM SIMPLE THINGS

I've been wondering whether my local hardware store, which seems to carry everything else, might have a product labelled "Universe in a Bottle." I would like to buy one of those, if only to read the label. Actually, I know what it would say: "Contents are 75% hydrogen and 25% helium. Trace contaminants less than 0.0003%. Consult a physician in case of accidental ingestion." There is not yet any bureaucracy that requires the acknowledgement of dark matter on the label. As for life, we don't merit special recognition among the trace contaminants!

I am intrigued by the universe's complex evolution from its simple beginning. Initially, the universe was as simple as anything can be, it was an almost uniform mass of quarks that quickly assembled themselves into an almost uniform mass of protons. Within minutes, they rearranged themselves to yield the specified portions of hydrogen and helium (with a trace of lithium) in an almost uniform distribution. You may already suspect that the adjective "almost" is significant. Those tiny fluctuations, which initially existed only on the quantum scale, were distributed by random gravitational forces in a manner that was not quite uniform. In fact, they were just non-uniform enough to seed stars and galaxy formation some millions of years later. Gravitational resonances brought about the collection of stellar dusts and debris into orbital zones in which stars and planets could take shape. On at least one planet circling one star, and perhaps elsewhere as well, conditions evolved that led to the formation of living objects. Along one of the branches that make up the bushy tree of life, there came to be entities with nervous systems. At least one of those species developed minds that influence and communicate, and that are derived from the functions of brains. And now we are perplexed by the complexity of our environments, our interactions, and even ourselves. Complexity overcame the simple universe by stealth, as a thief that came in the darkness of night.

I think that the concept of emergent entities is useful, but unfortunately not informative to frame an understanding of the foregoing. One cannot predict the properties of protons from their antecedent quarks—or of nuclei from that of protons. That is the nature of emergent properties. Similarly, each element differs from all others and from their millions of compounds whose unique characteristics do not overtly reveal their composition. As we extend the analogy into life, the basic pattern repeats itself. It is only through extensive analysis that life's relationships are revealed. Each one of our eight billion fellow humans has a claim on uniqueness at some point. We are all emergent entities and can claim a role for ourselves in the entire story of the universe. For the moment, it is a glorious and wonderfully complex vista that has been rolled out before us.

From this vantage point, what can be glimpsed about the future? About twenty five years ago, we discovered the dark energy that has our universe on an exponential expansion. In that model of the future, space between galaxies will continue to increase until the Milky Way and its gravitationally bound neighbours will be all that remains visible in our sky. In some models but not all, the expansion will continue until even the stars and planets are ripped apart and the black holes completely evaporated. Finally, after the subatomic particles have decayed, there will once again be nothing. As it was in the beginning, the universe will once again be simple in the end.

It sounds like an appropriate theme for a Beethoven's Tenth Symphony. Knowing that the entire universe, through matter, has been programmed for life and eventually also for death, one can feel a bit reconciled to one's personal prospects, knowing that we had a moment on stage in this cosmic drama.

HOW MATTER MATTERS

What happens when a gas is heated above 3,000C and thermal energy rips electrons away from their nuclei? Plasma is opaque to electromagnetic radiation. As an analogue, note that you can't see through a candle flame; for the same reason the early universe also was opaque. After the Big Bang, it took 380,000 years to cool and become transparent to visible light. Never mind that it took another one hundred million years before there was any light from newly forming stars. For the purposes of this line of thought, remember that in the beginning, matter was mostly in the form of radiation— it was very different from our current day to day experience of it.

That was the beginning. Now, let us cartwheel ourselves through spacetime and look at another four forms of matter of which we have absolutely no day-to-day experience, but which we now know are fundamental to the evolution of the universe. Put on your Cosmic Coroner's uniform and take a look at some aspects of dead stars.

Back in 1930, a brilliant high school graduate, Subramanian Chandrasekar (Chandra), was about to take up a scholarship to study astronomy at Cambridge University. In that era, the only mode of travel from India to England was a two-week voyage by steamship. How was a teenager to manage the boredom of the trip in the absence of any entertainment? This kid packed a ream of paper and a fistful of pencils, with which he set out to resolve how a dying star could collapse into becoming what we now call a white dwarf star. Upon arrival at Cambridge University, he announced that he had resolved the mystery by application of quantum mechanics. His mentor, Sir Arthur Eddington, set him up for bruising ridicule, but Chandra stuck by his evidence and was later awarded a Nobel Prize for work he had done on that ship before he was twenty years old.

So, what is a white dwarf star? In principle, it is the cinder of a star that was once about twice the mass of our Sun and that, in the process of dying, shed excess mass until it reached a residual mass of 1.44 times that of the Sun. When thermal energy from fusion was no longer sufficient to balance the force of gravity, then gravitational pressure collapsed the intra-atomic space that had previously been the domain of electrons. The result was that a non-compressible gas was formed, a form of matter for which we have no analogy on Earth, and that reduced the star to the size of Earth with a density of millions of tons per cubic centimetre. On the surface of this star, you would today weigh more than fifty tons!

It gets even more interesting in the case of larger dying stars. All of the above will happen as such stars collapse, but with the additional twist that the increased gravitational pressure will force the degenerate electrons directly into the atomic nuclei, where they will combine with protons to form neutrons. Neutron stars are much denser than white dwarf stars. A star with the mass of 10 suns (10 solar masses), will condense into a sphere of about ten kilometres in diameter (ie, the size of small city) and will weigh in at billions of tons per cubic centimetre. Neutron stars are structured with a unique core because they generate the strongest magnetic fields known in the universe—up to ten trillion times stronger than those on Earth! Not infrequently, neutron stars come in pairs that orbit around each other in decaying orbits until they collide. As their orbits get smaller, their spins accelerate to conserve momentum (as a skater who pulls in their arms to accelerate spin). As they approach collision, these super dense objects may be spinning up to thousands of rotations per second! These collisions are now known to be the main source from which most of the heavy metals in the universe were created. A typical such collision will generate a mass of gold and other heavy metals that is equivalent to one hundred Earths or more. The debris of such cataclysms has served as the source from which our Sun and planets were formed. If you are wearing any gold on your body, take a moment to acknowledge its exotic origin!

There is more and more weirdness to be had. Larger stars, exceeding ten solar masses, will collapse into black holes. Einstein postulated, and it's been proven since, that gravity is a distortion of spacetime in the presence of matter. In the case of black holes, the spacetime matrix completely collapses around them and cuts them off from any further contact with the rest of the universe. The density of matter in a black hole is infinite, and its dimensions are zero. The Schwarzschild radius is the theoretical radius inside from which there can be no return to the universe, not even at the speed of light. (Karl Schwarzschild deserves a footnote here; remarkably, he wrote a paper on the theory of black holes between artillery barrages while on a battlefield in WWI. Although he did not survive the conflict, his paper remains a landmark in the field of cosmology a century later.) Stephen Hawking postulated a quantum mechanical process of "evaporation" by which these bodies might again dissipate over the life of the universe. The process will take a long time, and it might be the last act of the universe.

Not a product of dead stars, but still incomprehensibly awesome, is the stuff known provisionally as dark matter. It's been in the universe since the beginning of time, but we only noticed it recently. Fritz Zwicky noticed about seventy years ago that given their high orbital velocities, galaxies ought to be flying apart. He postulated the existence of "dunkelous materia" to account for the orbital stability of stellar systems and the galaxy. If he hadn't been such a difficult character, he might have been listened to, and we would be a generation further in understanding what he was referring to.

It was left to Vera C. Rubin to make the required observations and develop the underlying theory.

What we do know today is that in gravitational terms, there is at least six times more dark matter than normal stuff. Dark matter envelops the galaxies like an escort who maintains the orderliness of our convoy through space; without it, no galaxies would ever have formed. At the same time, we have no idea what it may be. We hypothesize that it is a subatomic particle that does not react with light or matter, except through its effect on gravity.

Normal matter and dark matter together account for just thirty percent of the mass of the universe. The remaining seventy percent bears the placeholder name of dark energy. Dark energy seems to be a feature of space but we have no evidence to say what it is.

It was wonderful to discover the plasma state of matter, because it explained a lot about the early state of the universe. Just as awesome are the states of matter that mark the deaths of stars and the mysterious dark matter that holds it all together and spreads it apart. At the same time, I have renewed respect for the common forms of ordinary matter, of which bedposts, broomsticks, and I are composed. The thought that we are all a part of this is at least as moving as a full-throated symphony.

HOW MUCH FOR A
BAR OF SOAP?

I'm thinking that the invention of soap must have played a major role in the development of civilization. A product of Middle Eastern science in the time of the European Dark Ages, soap became a valuable item in the Western trade for spices and other exotic products from the Orient. I can imagine a curious native being lured into a Florentine back alleyway by an accented foreigner with a couple of loaded camels, offering a bar of soap for sale along with instructions on how to enjoy the first bath of his/her life. Better than a drug-induced high!

It was Arab chemists who discovered the alkaline properties of the leachate from wood ashes; to this day, our word for the involved elements, "alkali," is the Arabic word for "ashes." Many more important discoveries were made during the Islamic Golden Age, but more on that another time.

My mind was dragged back to this line of thought by a recent radio conversation about the challenges of living in a nuclear fallout shelter (or any other tightly enclosed space, such as a rocket ship) for an extended time. One facet of the problem was the challenge of maintaining personal hygiene under the expected conditions. One guest suggested that in the event of nuclear war, the obvious solution would be to make one's own soap right there in the shelter. Right! I Googled a current soap recipe: 2/3 cup of coconut palm oil; 2/3 cup of olive oil; 1/2 cup of sunflower oil; 1/4 cup of NaOH (lye). Moreover, the recipe also called for a 1/2 cup of dried flower petals for aroma! Just the range of ingredients one is likely to have on a shelf at the back of the cave after several weeks of sheltering!

I learned soap making from my mother before I went to school in rural Saskatchewan. It involved rendering the tallow of a cow we had killed in the fall for our winter meat. To this day, I could skin an animal and flense the fat from the hide and carcass for the production of soap and candles. I remember how the tallow was carefully heated in a large boiler until all of it was liquid, and the cellular debris had settled in the aqueous phase at the bottom of the pot. I also remember that I was banished to the extreme far corner of the room while Mother ever-so-carefully added in the lye. This step involved a highly exothermic reaction, and it could go badly wrong—explosively so. In this manner, we obtained a year's supply of soap, such as it was, for a cash outlay of a single can of lye, some ten to fifteen cents. This was a rough and universal soap that we used on our bodies, as well as the dishes, floors, and the laundry. It wrecked our hair and left our knuckles raw and itchy, but thanks to Mother's oft-repeated mantra that "cleanliness was next to godliness," we were at least clean. The idea that soap should also have a pleasing aroma? Unheard of! The thought never occurred to us; wasn't it enough to not stink?

I suspect that when our hypothetical nuclear war survivors have exhausted their supplies and resorted to chewing their boots for dietary protein while waiting for ambient radiation levels to decline to safe levels, they will no longer insist on exotic oils in their vegan soaps. They may even learn to be simply grateful for the ability to still scratch their itchy, unwashed bits. Where will the travelling soap salesman be when we next need him?

HOW SWEET IT IS

Four centuries ago, Paracelsus pronounced that everything in the universe is toxic at the proper dose. I think of him often, and especially when self-acclaimed dietary experts publish books about their nutritional fixations and then expound on them in the media. At the proper doses, even oxygen, water, and glucose are lethal, although in proper measure they are each essential to life. Let us take a few moments to consider the role of glucose in our lives and, more broadly, in the biosphere.

As a biochemist, I take great pleasure in contemplating the pathways delineated on a metabolic chart. At the centre of the action is always the sugar (ie, glucose), which can be metabolized without oxygen to lactic acid and alcohol; processed into acetyl fragments for oxidative breakdown to carbon dioxide and water; retained in reservoirs as muscle and liver glycogen against short-term future needs; or processed into fats for long-term storage. All of our tissues can use glucose directly, and the brain is exclusively reliant on a second-by-second supply of both glucose and oxygen for its continued functioning. Our problem with glucose and other sugars is that we like them, and excesses lead to obesity, type II diabetes, and other life-shortening problems. For our ancestors, whose life expectancies were shorter, a tendency toward obesity was probably a good thing, as it would have provided an extra margin in times of starvation. In any event, they would rarely have lived long enough to experience its adverse effects in older age. In this way, evolution in our ancestors selectively preserved genes that favour obesity.

Glycogen is a branch-chained polymer of glucose that forms in liver and muscle cells when more glucose is present in the blood than we immediately need, and it is the reservoir from which we regulate our energy needs on a moment-by-moment basis in stressful situations. Glycogen stores are a source of the "second wind" that athletes experience in long duration events. There are about a half dozen diseases in which the breakdown of glycogen is prevented by specific enzyme deficiencies, and in these instances, glycogen deposition is not reversible, and these stores are then not available when needed.

Another form of glucose storage is found in plants in the form of starch. Starches resemble glycogen, but with less branching in their molecular structure. Plants use starch as the preferred food source for their embryos and their energy source at night. Grains (e.g., wheat, oats, barley, rice, corn, etc.) are among the chief sources of starches (carbohydrates) in the human diet; and they are all glucose polymers.

We don't often think about sugars when we are hiking through wooded areas, except perhaps for a nod in the direction of a sugar maple stand or a restorative break with a candy bar. But stop to consider that the mechanical strength of trees, corn stalks, and wheat straw are mainly due to cellulose. And cellulose is another glucose polymer. It differs from starches in the matter of the angle of the C1-C4 inter-glucose bonds, which makes the retrieval of single glucose monomers by enzymatic breakdown difficult from cellulose. That is why logs on the forest floor take years to rot away and do not dissolve appreciably in a summer rainstorm. Our wooden-framed houses are essentially a variant of gingerbread, except that they do not dissolve in the rain. However, cattle and termites harbour organisms capable of breaking down celluloses in their gut, which enables them to benefit from the ingestion of grass or your floor joists, respectively.

While they may become items on future menus, insects rarely come to mind when we are talking of sweet foods. Insects characteristically have an external skeleton made of a polymer of a modified glucose molecule known as glucosamine, which is glucose to which an amino chemical group (nitrogen) has been added. The world biomass of insects may exceed that of all other life, and the amount of glucosamine contained in their chitin is correspondingly large.

I have just identified more latent glucose than we will ever need to maintain our basic functions. Where did all this excessive supply come from? Is it another facet of the conspiratorial sugar industry to submerge us in a gross dietary excess? Actually, all of it comes to us as a gift from the Sun. The process of photosynthesis in green plants captures carbon dioxide from the atmosphere and adds it to a five-carbon sugar (ribose) to make glucose and thereby takes millions of tons of CO_2 out of circulation every year.

There is one other fascinating aspect to glucose; it comes in two forms that are chemically identical in terms of their reactivity but are metabolically exclusive of each other. When glucose is synthesized in the laboratory, the product always consists of an equal mixture of two molecular forms (isomers) that resemble each other in the way that the left hand resembles the right. We call them optical isomers because when separated, they rotate a beam of polarized light to the right (D) or the left (L) respectively. In contrast to the lab synthesis, all the glucose produced or used by all life is of the D configuration. This observation is consistent with the notion that all life is related through a common origin; had there been many origins, there should have been some that, by chance, were centred on L glucose as well.

There are other sugars beside glucose with key functions in life. Ribose, the five-carbon sugar mentioned above, is a fundamental component of DNA; each human cell requires more than six billion molecules of a ribose variant, two-deoxy-ribose for the structure of its chromosomes. Similar amounts of L-ribose are needed to build and maintain the messenger RNA, ribosomal RNA, and amino acid-specific RNAs that are required to transform information coded on DNA into proteins.

So, here we are in real life, surrounded by the mystery of sugar in quantities that dwarf anything that we ever imagined. Glucose is the energy source that powers our bodies and, especially, our minds. In the wider totality of nature, glucose stored as starches holds energy in reserve for future needs, and it also makes a durable structural cellulose scaffolding that allows trees to stand on the shoulders of their former selves, year by year, while insects draped in protective chitins, also essentially made of glucose, scavenge their edible bits. More and other sugars are needed to maintain the integrity of our genetic details. Who would have thought it possible? In these doses, sugar is not so much toxic as it is joyfully intoxicating. Paracelsus would have loved this scene.

I AM FEARFULLY AND WONDERFULLY MADE

As important as they are to our economy, I do not like television or radio commercials. Mostly, they are simply annoying—somewhat like a persistent fly about my nose. I resist the intent of the advertisement, which is to implant a persisting brainworm, as much as I can. Nonetheless, resistance can prove futile, as it was in the case of the candy bar ad with the tagline, "You're not you when you're hungry." Then it came to me that a shorter but similar statement could be used to convey a still deeper truth: "You're not you." We simply are not who we think we are physically, and that is true in at least three ways, which have implications for how we think of ourselves in the world.

When I was in medical school, more than sixty years ago, the curriculum for the first two years was largely devoted to mastery of anatomy and physiology. To this end, we spent something like 400 hours dissecting a body and countless hundreds more hours viewing the microscopic features of all body tissues in both healthy and diseased states. One of the sub-courses of those years was microbiology, which generally featured microbes as the enemy to be shot on sight. Antibiotics were new then, and it made us feel so powerful to imagine we could kill every bacterium we saw to the betterment of our patients. There was a perfunctory nod given to the 'good' microbes, such as those that we use to make bread, beer, and yogurt, but nothing more was made of those. In the years that followed, we have become progressively more aware of the intricate relationships our bodies have with the myriad of bacteria without which we now know we could not exist.

That's right; we could not exist without our bacteria! Since bacteria are much smaller than typical human cells, they occupy less space; however, if we put them all together, they would form a mass larger than a normal liver, making them the largest organ in a healthy body.

One way to look at this situation is that if our bodies were organized along corporate lines, then we would be minority shareholders at every meeting, and we would need to collaborate with bacteria in order to survive; and we do. We missed all this back in medical school only a short while ago. If one thinks about this for too long, there is no avoiding the question, "Who is in charge?" Is it plausible that bacterial urban planners invented us as essential infrastructure to support their own microbial civilizations? Do they affect who we are? Certainly, it is true that some microbes can and do change the behaviour of their hosts. For example, rats infected with toxoplasma lose their fear of cats, which is bad for rats, good for toxoplasma, and unimportant for cats beyond an easy dinner. We are on the lookout for evidence that such links may exist between us and our microbes.

At a second level, it gets much worse than just that. Our information systems have been hacked. This finding is comparable to the problem that Sony had with the North Korean hackers a while ago. The essential information that we need to build and maintain our bodies is encoded in DNA, a linear molecule consisting of more than three billion units called bases, packed into forty-six chromosomes with one complete set in each cell. This DNA, which we pass on to our children, has been hacked and overwritten with bacterial and viral messages in about 23,000 places, which totals about ten percent of our genome. Has any of this microbial graffiti altered our genes?

What do we owe to these interlopers for their editorial input into the nature of our existence? Is any unique human attribute due to their intervention? As it happens, the answer is resoundingly, "Yes!" Just one of those viral genes from long ago encodes the protein syncytin, which makes placentae possible and may be the item that converted egg-layers into mammals.

There is yet a third way in which we are not quite ourselves. Anyone who has ever managed an industrial warehouse knows that it is costly to keep obsolete materials in stock, and there is pressure from management to clear out everything that is not contributing to the company's current productivity. Our bodies may, in one sense, be compared to grossly overstocked warehouses without proper inventory controls; or, if you wish, they may be regarded as museums with very high operating costs. There are several hundred genes that we no longer use, which once operated our sense of smell (and still do in animals such as dogs). These genes are still present, but they have been switched off, and they cannot be turned on again. We also have all the genes necessary to make our own vitamin C, as other mammals do (except apes and guinea pigs), but they have also been irreversibly switched off. Nevertheless, we must pay the metabolic cost of maintaining them in their present useless state.

Where does this leave us? I am not the controlling shareholder in my own body, my intellectual property has been hacked, and I am seemingly stuck with the maintenance of an expensive store of historically interesting, but now selectively useless, inventory that I don't know how to get rid of. Three thousand years ago, a poet said, "I am fearfully and wonderfully made." He didn't know the half of it, and I suspect we don't either.

I WANT TO BE AN
ASTROBIOLOGIST TOO

When Edwin Hubble was a teenager in the American Midwest, he was fascinated by the stars and told his family he wanted to become an astronomer, to which his father responded, "It's all well and good to be interested in such esoterica, but my kid has to qualify for a line of work that pays the bills." He then sent Edwin off to law school. Being an obedient son, Edwin went. He did well and even won a prestigious Rhodes Scholarship, which allowed him to study in England for a couple of years and become a first-rate basketball player. On his return home, he set up an office law practice in his hometown. Then his father died. No sooner was father safely buried that Edwin announced his retirement from law, closed the office, and took a job teaching high school science while he applied for graduate school programs in astronomy. I don't believe he ever had any difficulty paying his bills thereafter, and he became the best-known astronomer of the twentieth century for his discovery of the expanding universe.

Turning to the present, how will you respond if, at dinner, one of your children announces an ambition to become an astrobiologist? You hear the word, but what does it even mean? How can astrobiology possibly pay the bills? Where should one look for life in space? Who wants to know, and what's that worth? To that end, we would do well to first closely study the one example of life that we have on Earth.

How easy/difficult might it be to recognize exotic life forms around us, never mind something that might turn up on another planet? Fifty years ago, when I took up beekeeping, there were a couple of entomologists (the Goulds) who felt that to prepare for the recognition of alien intelligent life, we should first study the many different forms of possibly intelligent life on Earth. They chose to study honeybees. In one experiment, a food source was placed a metre from the hive at 9:00 a.m. The next day, the source was placed at two metres, and then successively at four, eight, sixteen, and thirty-two metres, always at 9:00 a.m. The day after thirty-two meters, the experimenter delayed placing the feed until 10:00 o'clock and he found the bees waiting for him at sixty-four metres! It seems that the hive, perhaps even the individual bees, can resolve an exponential function while also keeping time. What else have we missed about other intelligences surrounding us on Earth? Short of anything we might call intelligent, there are other mind-clotting surprises in store here; consider the creatures that require a bath of boiling battery acid to survive, or those that thrive on the lips of undersea volcanoes at 122°C, or the bacteria I once studied that can survive three million rads of ionizing radiation. To reiterate, the first thing an astrobiologist needs before sallying forth into the wider universe to spot exotic life is a deep reference understanding of life on Earth.

The second set of things for an astrobiologist to know is about other planetary systems. Although the study of Mars has not yet yielded any signs of life, we have not completely given up on that possibility because our data so far don't tell us about more than the first few centimetres of the surface. There might be anaerobic biota lurking deep in the belly of that planet. The biomass of life in our Earth exceeds that of all the life on the surface. The biomass in the oceans dwarfs surface life by hundreds, perhaps thousands, of times;

we don't yet know about the state of affairs on Mars. We may need to drill deep to know.

Going beyond our own solar system, as recently as twenty-five years ago, no one had ever seen a planet orbiting another star, and we didn't know if there were any. The Kepler Space Telescope project then identified more than 5,000 exoplanets on a sample of just 150,000 stars. Planetary features that might permit life, as we think of it, seem to exist on some of them. In a statistical way, we know that there are many more exoplanets than stars in the sky—more than one hundred billion in the Milky Way alone. The Kepler Project has been completed and is now carried on by its successor, named TESS (Transiting Exoplanet Survey Satellite), whose design allows monitoring of the entire sky. Which of these many exoplanets are candidates for life? Even if it is only one in a million, then there are still a hundred thousand candidates! We can assess for the so-called Goldilocks conditions—ie, liquid water and an atmosphere—by use of spectroscopy. Transmission spectroscopy performed by powerful telescopes, such as the James Webb space telescope, will provide this information.

To come full circle and back to Earth, the identification of candidate planets for life urgently raises questions about how life began here. Since all life on Earth is composed of the same elements, and since we know that exoplanets are composed of the same abundances of elements as Earth, it seems plausible that the development of exolife will have taken similar paths. This line of thought has raised many, suddenly urgent questions for astrobiologists about our own beginnings. How did we get from an inorganic planet to become one with life? We need to have a deeper knowledge of our own earliest beginnings to do justice to whatever comes to us through our wonderful new telescopes. Current hypotheses concerning prebiotic chemistry favour the appearance of chemical reactions on the surfaces of rock crystals, presaging the evolution of enzymes. The presence of mineral atoms at the active centres of many of our enzymes is taken by some to represent a variety of molecular fossils residing in modern life.

I envy the astrobiologists who are so boldly facing a bewildering and wonderful new frontier; I think parents needn't worry about how their kids will pay their bills and they may not have to drill for oil on Mars to hit paydirt.

IF ONLY IT WAS TRUE

The Missionaries of Good Nutrition these days are warning us of the harm that will come unless we take steps to inactivate the free radicals in our diet. They allege that certain foods are more likely to contain these free radicals, and that some other foods, dubbed "free radical scavengers," contain their antidotes. The solution is to eat the right things. Colour me deeply skeptical.

First, I will say that there are a lot of publications relating to diet and nutrition. Like most good research, dietary research is hard work to do properly according to modern standards. Gone are the days of questionnaires (ie, how many times per week do you eat potatoes; how many helpings?). However, a lot of so-called nutrition research reduces to self-promotion or ideology, often the kind of research that sets out to validate a predetermined perspective or decorate mere imaginative speculations with words like "clinically proven" in their adverts rather than any tests of reality. I will limit my remarks here to the biochemical issues related to the metabolism of free radicals and mitochondria, of which I may know something professionally.

Free radicals are unstable molecular species that have picked up an unpaired electron as glucose fragments pass through the oxidative metabolic chain located on the inner membrane of the mitochondria. Think of them as energy production accidents that fell off a hand grenade assembly line on the way to becoming ATP; you shouldn't drop a grenade in assembly either. Because they are so very reactive, free radicals tend to interact irreversibly with the first molecule they encounter and, most often, that is a water molecule. Water, in turn, passes the charge on to an oxygen atom. Free radical production accidents are not happening everywhere at random in the volume of the cell; further, their migration from the cell is not unimpeded. First, they need to pass through several membranes that have a high probability of stopping them: the inner and outer mitochondrial membranes and the space between them. Next in order is the cytoplasm, another space that is heavily infiltrated by a membrane structure named the endoplasmic reticulum, and then the cell membrane itself. While this is going on, the free radical is undergoing exponential decay with a half-life of about a second or two.

Now consider the blueberries you ate with your cereal this morning and the virtuous free radical scavengers they supposedly provided. These also followed an obstacle-laden path on their way to do battle with the free radicals. First, they encountered the proteolytic acid bath of the stomach, which is itself a strong electron sink, and that was followed by the alkaline experience in the small intestine, by which protein fragments were reduced to their constituent amino acids. That was followed by the microbial bath in the lower small bowel, where there were sure to have been lots of free radicals generated by microbial metabolism. Whatever free radical scavenger molecules remained after this point, then had to traverse the multicellular layers of the gut and enter capillaries in the blood stream and tissues. I haven't found any evidence for or against the probability that the free radicals or their scavenger molecules survive to meet each other from the two ends of this journey.

I've listened to several unconvincing discussions concerning the supposedly wonderful protective antioxidant properties of selected foods; I think it is all a naturopathic marketing ploy. The enthusiasts for the concept maintain that free radical scavengers slow down the rate of mitochondrial aging and our overall aging processes, but the evidence-based literature clearly shows that mitochondria are normally continuously replaced by newly synthesized copies, independent of the age of the cell, and their aged-out constituents are continuously recycled by a process called mitocide. Nature has evolved mechanisms that take care of mitochondrial aging without regard to our diets. It would be so great if one could eat one's way back to youthfulness through an orgy of blueberries; that would be an ultimate wish fulfillment. If I have any advice to give, it is this: eat as many blueberries or other so-called free radical scavengers as you can enjoy, and don't let anyone seduce you into thinking of good food as mere medicine.

IN THE FULLNESS
OF THE SENSES

Last night it snowed a bit. Not a lot. Just one of those cautious, stealthy snowfalls that happens, silent as a breath, while no one is looking—the kind that lodges itself on every branch until it has clapped a sonic filter over the entire scene through which I take my morning walk. Lost as I was amid the wonder of the setting, I realized that silence can be so much more than the mere absence of noise; rather, it can be something analogous to the way that space is proving to be so much more than just the emptiness that separates objects, but a substantial medium with its own set of parameters and characteristics.

I believe we should take full advantage of every opportunity to live with fully enabled senses. Eyeglasses, hearing aids—I'll take what's available. Our collective imperfections have enabled entire industries in optics and audiometry. At the same time, I insist to be allowed to take them off when I'm unconvinced that I need to understand everything that is going on before me. I wear earmuffs when engaged with the noisy aspects of my woodworking hobby, and sometimes it is comforting to leave them on to enhance the silence that follows.

The communities of the blind and the deaf often argue that the rest of us designate them as disabled when they are only differently abled. I thought of that last evening as I watched the blind Canadian skier at the Paralympics who fell early in the race but then progressively overtook the entire field and finished in first place. I recall a recent video clip of a blind cyclist who rides his bicycle in city parks and avoids trees by the echo-location techniques that his blindness has taught him—yes, just like a bat. More than fifty years ago, our local paper featured a blind cabinetmaker who could cut a board on his twelve-inch radial-arm saw to the exact required dimensions, and who claimed to avoid injury by sensing the movement of air on his fingers near to the whirling blade. An eye specialist eventually looked in his eyes and announced that something could be done to restore his vision. In his postoperative interview, the cabinetmaker somewhat ruefully remarked that sight wasn't all it was made out to be: "Now that I can see my sawblade, it scares the hell out of me." These differently abled folks may be onto something, and I wonder what the sighted and the hearing are missing.

Until now we have used technology to make up for shortcomings in our vision and hearing. What if we could move on to augmentation? In a way, we already have; military night vision goggles are an example. Would you accept a further innovation in eyeglasses that provided a rear view or that resolved visual limitations due to our blindspot.? What impact might there be from extending our visual spectrum into the ultraviolet like birds and insects? We are doing that with the James Webb Space Telescope, which gives us views of the sky in the infra-red spectrum and allows us to peer through dust. What if we could experience new senses, like migrating birds who apparently experience magnetic fields through their beaks? How might we integrate bird beak sensoria with our GPS technology? The mind boggles at the borders of the unexplored, and who can say what is coming next?

INCOHERENCE ADDRESSING THE INCOMPREHENSIBLE

Quantum mechanics has always been a step beyond my pay grade, possibly because I don't have the vocabulary to sustain a coherent line of thought on the subject. Lately, I noticed that some other folks are not deterred on this account. To be charitable, perhaps quantum mechanics is incapable of being fully expressed in spoken language. It takes a lot of metaphorical flag-waving (and some difficult math) to impart the essential concepts, and even after the effort has been made, someone who disagrees will be found.

Most such discussions begin with references to the double-slit experiment, or to a version of it that uses interference spectrometry. If you missed the reference, recall the pinhole camera of high school physics and invent an advanced camera with two pinholes; then stretch the pinholes into slits. To my ear, the incoherence of opinions happens, not in the experimentation, but in the ensuing interpretations: is matter coherent at the atomic level, like bullets, with mass, velocity, and trajectory? Or is it diffuse, with force, amplitude, and relativistic velocities? The answer seems to be affirmative, and it depends on the presence of an observer to select the difference.

When the universe was young (i.e., during the trillionth of a trillionth of a trillionth of the first second of its existence), it was riddled with quantum uncertainties—the tiniest variations in proton densities that ought to have been corrected through random diffusion in the next instant of time—except that it abruptly underwent inflation at a rate that exceeded the speed of light by about sixty orders of magnitude in a trillionth of a trillionth of a millionth of a second. This was not an instance of matter being exploded into space, but of space unfolding and carrying matter with it at velocities that exceeded the speed of light. In this way, regions of space with those small quantum irregularities were irreversibly separated from each other forever. Those irregularities of the nascent universe became baked into our present-day reality, and that laid the background from which the universe evolved into its present form. I suspect that if "the inflation experiment" could be repeated an infinite number of times, each result would be unique though similar to the others.

I am reading Kevin J. Mitchell's book, *Innate: How the Wiring of our Brains Shapes Who We Are.* The book offers a description of the molecular, genetic, and morphologic development of the human brain, and I realize that this fantastic process is not so different from the evolution of the early universe (not that it necessarily is quantum mechanical in nature). Brain development also requires an immense cascade of events that promise uncertain variations of outcomes each time to produce yet another unique personality. From a single cell whose information is encoded on 3.2 billion nucleotides along forty-six chromosomes, an individual is formed, the like of whom has never existed before, and this has been the case for more than ten billion iterations, since humanity first emerged as a distinct species. Even identical twins have their unique aspects, beginning with about one hundred random mutations after their first cell division. That same small replication error rate continues with every replication and is enhanced by response to micro-variations caused by epigenetic influences that impact on the cells unique to each embryo.

Roger Penrose and a few others have attempted to invoke quantum mechanics in the development of consciousness, but with uncertain success. While quantum mechanics surely dominates the subatomic behaviours of our constituent molecules, I fail to see how it can play a role at the scale of the living body. Not every bit of extreme complexity is explicable by reference to quantum mechanics. As I said at the beginning, I'm not sure that adequate language has yet been invented to permit a proper discussion to develop. Nevertheless, we must keep on trying.

IS BIOLOGY STILL DISGUSTING?

Back in the mid 1950s, the students' newspaper at my university ran a riddle:

> *How can you tell which science lab you are in?*
> Answer: *If it stinks, it's chemistry; if it doesn't work, it's physics; and if it's disgusting, it's biology.*

Back then, biological museums were filled, as they still may be, with butterfly, bird, and fossil collections, which "proper scientists" (aka chemists and physicists) dismissed contemptuously as "stamp collections." If your memory goes back to those times, then the discussions that go on these days around vaccines must be truly mind-bending, since biology has become as atomically precise and mechanical as anything that chemists or engineers could develop.

We are now in the era of plug-and-play biology, and advances in understanding and innovation are occurring at a dizzying and accelerating speed. The Internet now carries photos of cats that have incorporated a jelly fish gene, which causes them to glow in the dark—a feat that required moving a specific fragment of DNA from a jelly fish chromosome into the germ cell of a cat. I suspect that for mice, this is as good as a bell.

Other examples of cross-species gene transfers that come to mind include human insulin, now routinely grown with instructions from a human gene transplanted into *E. coli* bacteria for almost all the

147

world's insulin production. Similarly, human thyroid stimulating hormone is produced in cells from a Chinese hamster's ovary with the appropriate human gene inserted and turned on. This is truly the era of plug-and-play biology. And it's not difficult to do: the essential ingredients are largely available through Amazon, without regard to qualifications of the purchaser. High school science fairs will show evidence of that. Similar resources have also enabled a very adroit response to the vaccine-building challenge posed to medical science by the coronavirus strains in recent years.

Let's consider the timeline of that crisis. On December 20th, 2020, the world learned that a pandemic was coming, and the experts knew from their viral surveillance that the agent was a specific coronavirus, subsequently named Covid-19. On that day, work began to determine its genetic (RNA) sequence, and that work was completed on January 10th, just twenty-one days later. Over the following three days, there were urgent meetings involving scientists, pharmaceutical companies, and governments from around the world, in which all the options for the creation of a vaccine were explored in broad discussions. Three approaches emerged: make an antibody to the protein on the spikes on the outside of the virus; target the viral cell wall at other sites; or attack the viral RNA directly. The latter was clearly the newbie dark horse in this competition, and it seemed most likely to fail. Different companies elected different strategies, and from there, it became a race to the finish. The dark horse won the race against long pretest odds.

The recipe for the successful vaccine includes a piece of viral RNA enclosed in a fatty membrane and that, in turn, codes for the spike proteins; the viral RNA directs human cells to make spike proteins, and these in their turn direct the immune system to make the antibody. The antibodies produced are even more effective than the models predicted. In even the recent past (ie, the last twenty years), this body of work would have taken at least a decade to complete. The conspiracy theorists who believed that the rapid pace of apparent success was concealing shortcuts were right about one thing: there were shortcuts taken to get to the final product—not in the sense of shoddy workmanship, but in anticipation of the supply chain requirements and inventory that full-scale manufacture would require as soon as the experimental data permitted. Before they knew for certain that the vaccine would work, the main players had placed millions of dollars at risk, but they were also confident that their approach was the correct one. When they were proven to have speculated correctly, they were ready immediately to go into production.

I argue that biology has passed the smell test. As miserable as these recent years were, they did not wreak as much havoc proportionately as did the flu pandemic of 1917–9, when an estimated fifty to one hundred million people died out of a world population of then only two billion. Thanks for the limitation on casualties goes to the biologists and molecular scientists, and to the scientific and health care support systems, whose curiosity-driven research over decades supported the vigorous vaccine developments that gave us hope in a dark time.

NOTE ADDED IN PROOF: As I edit this essay, I am aware of an impending possibility of an oncoming epidemic of bird flu. The experts are saying in this week's news (June 20th, 2024) that the virus is two mutations away from releasing itself into the public and spreading among us later this summer. Although nothing is being said, I am confident that the vaccine producers are already at work to prepare the next vaccine we will need. Readers will know if that became necessary.

IS IT A TREATABLE ILLNESS?

Isaac Asimov was an American science writer and professor of biochemistry, but he gave up on laboratory science at an early age because he loved writing books more. Overall, he produced more than five hundred tomes. Today, he is mostly remembered for his elegant science fiction novels.

One of his early fiction works was *Nightfall*. It concerned an inhabited far-away planet, named Labash, that was illuminated by a cluster of six stars such that no part of the planet was ever in complete darkness. Creatures there had no experience with cycles of light and dark, or even of darkness proper. Their psychologists were intrigued by the observation that merely placing experimental subjects in a closed windowless room caused them to suffer hysterical attacks that were only sometimes relieved upon their return to light. Then their archeologists discovered evidence of previously unknown civilizations that seemed to have all ended in obliteration by massive fires. Further studies then dated these cycles of fire as having recurred at regular intervals of 2049 years.

And then it happened without any warning that they could have recognized beforehand. At first, the sky became dimmer than usual for that time of day. As the dimness progressed, people began to become vaguely uneasy; some thought to ease their anxiety by setting bonfires in the backyard firebox with their families, but the darkness progressed further and became frankly eerie; more light was needed.

At first, it sufficed to burn scrap wood that was in the backyard and in their garages; the doghouse also worked. Then, some intrepid soul looked up into the dark sky and saw stars for the very first time. The hysteria was instantly infectious, and the entire neighbourhood went berserk. No one had ever seen stars before, nor had they any context to support what they saw. There was no telling what would come next.

More light was needed to restore perspective. First it was the vacant house on the corner lot that was set alight; then, like a spreading plague, it was the neighbourhood; then it was the downtown district. When the light did return and the suns again emerged with their usual brilliance, the total devastation became evident. Scholars recalled the layers of ash, and then they guessed at what had taken place. "We told you that something like this happened in the past, and it will now take us several thousand years to recover to where we were at only yesterday. We must master our fear of the dark if our descendants are to survive the next such terrorizing phenomenon," they said. But their communications media had also been destroyed, and no one heard them.

Let's listen in on an astrobiology lecture at the University of Labash in the year 2020. Their last trauma of the Great Darkness is several millennia in the past, and few people ever think of it anymore. There is some knowledge of stars. Scientists are excited by recent contact coming from a civilization of beings on a planet that calls itself Earth. The people there are struggling severely with their own terror, a plague that is killing millions of them by obstructing their ability to breathe. The humans know that prevention is possible if they isolate themselves from others and wear masks, but they are social creatures, and in increasing numbers, they are refusing to comply with public health advice. Instead, they are blaming their leaders for fostering a false fright intended to perpetuate limitations on their freedoms and tighten controls on them by requiring them to isolate and wear masks. Clearly, these unfortunate humans are afflicted by a brainworm that is compromising their ability to reason clearly. The plague is treatable; the brainworm, not certainly so.

The lecturer is contrasting the Labashian instinctive terror of the dark and the Earthlings' apparent fear of masks. What is to be learned from the contrast? Are we all, Labashians and Earthlings, untreatable victims of external forces? Would either species change their behaviour if they understood the causes and principles of social support and management more clearly?

As she is wrapping up her lecture, the professor notices an odd sense of dread crawling up from the back of her mind. The class is atypically quiet as the students file out of the room. The scene outside the windows of the lecture hall is quickly and uncomfortably darkening. And then, a terror-stricken voice shouts from the corridor, "Hey. The shopping mall is on fire."

IS SCIENCE EVER USELESS?

I'm sure you've sometimes wondered about the wisdom of allowing scientists to follow their unbridled curiosity, especially if the cost seemed to be the most awesome aspect of the project, or the outcome seems to have no immediate utility. Surely it occurred to you that a more beneficial expenditure of our scarce resources would have been possible. The nearly 10 billion dollars (adjusted to 2024 currency value) it cost to order the sequence of the 3.2 billion bases in human DNA two decades ago might have seemed like such an example, and a committee of the American Congress did ask that question a decade later. Their committee (circa 2006) found an accelerating economic gain, much of it for new and better medical diagnoses and pharmaceuticals, of $140 for every dollar spent on that project; and the gains continue today. The benefits of a project that might have seemed pointless to some at the outset continue to exceed expectations.

Consider the field of forensics. In the 1960s, a nurse, Gail Miller, on her way to work at the Saskatoon City Hospital, where I trained, was murdered and her body was found in a back alley. The investigators suspected David Milgaard, a local teenager and he was charged with the crime. David may have been undergoing a manic episode, and his behaviour around that time drew the attention of the community. However, despite his protestations of innocence, the court found him guilty of murder, and he would have been forgotten except that his mother believed in him and kept his case before society for many long years. Eventually, the methodology of DNA sequencing became affordable to crime labs, and David's DNA did not match the specimen that had been retained from the scene. The true criminal turned out to have been hiding in plain sight; his DNA was a match and he had previously served time for assaulting women. David was released after missing twenty-two years in which he ought to have been making a family, building career, and developing his rightful place in society. Thereafter, he championed the cases of the wrongfully convicted until his death in 2023.

I kept casual track of similar cases of wrongful convictions in Canada for a while, but they were soon too many to remember. Each had its own toll of years lost and lives damaged, and many of them were later exonerated by the new technology. In our time, DNA testing is the work of a few days at most; innocent suspects of such crimes are quickly screened out, and I've not yet heard of a false-positive identification with DNA. Forensic applications are among the many spinoffs of this bit of basic science.

Also, consider the cost of sequencing ancient DNA from archeological specimens, with all the unique problems relating to degraded materials and intrusion of foreign nucleic acids. The complexity of the problem is comparable to solving a billion-piece jigsaw puzzle that has been contaminated by several thousands of incomplete thousand-piece puzzles! Nevertheless, billions of dollars have been spent to construct and support special labs for the purpose, and now we can tell scientifically informed bad jokes about the Neanderthals lurking within us. Did this work have an application? All people of European background are about two percent Neanderthal. Each of us has a different two percent, and if it were all added up, it would be true that there is more than fifty percent of a Neanderthal at every major sports event. But surely, we didn't analyze Neanderthal genomes just for this weak amusement! We have learned, to our amazement, that we owe much of our cellular immune system to these ancient cousins. Think of that when someone around you next needs an organ transplant.

In summary, at the front end we have to take some research proposals on faith. Don't let anyone tell you that the billion dollars spent to set Percy down on Mars, plus another billion ten years hence to bring those Martian soil cores home, is a gross misspending of public funds. The project is sure to teach us something, although I can't guess yet what that will be. We just have to stay up for the whole movie.

LIFE'S LEGO BLOCKS
(AND MORE?)

By way of introduction, I need to speak of the earliest living forms that we know of. First on stage seems to have been a cast of life forms we call archaea, microscopic forms that live without oxygen and have established a foothold on and deep into the Earth. We only learned of their existence in recent decades. Because they look like bacteria under a microscope, we did not appreciate their distinctiveness until we could analyze DNA sequences. Archaea and most bacteria will only grow in their natural communities, but not in the laboratory. Their presence can then only be recognized by finding their DNA sequences in earth samples. However, they dominate the planet. Bacteria seem to have evolved from archaea; among these were the cyanobacteria, who first used sunlight for energy and excreted oxygen into the atmosphere. Other bacteria then developed metabolic pathways that enabled them to use oxygen.

It is a feature of both archaea and bacteria that their cells are small in comparison to the cells of later-occurring plants and animals. This is because they had limited energy resources. Then, a couple of billion years ago, an archaean and a bacterium agreed to cooperate and, in so doing, they formed the first eukaryotes. If this was to be a successful venture, there had to have been an amicable division of duties. To this end, the archaean component took control of DNA synthesis and many details of metabolic management in the nucleus, while the bacterium concentrated on the adaptations that led to the efficient use of oxidative metabolism to produce energy. In fact, the same glucose substrate now produced thirty-eight molecules of energy-rich ATP per mole, compared to the previous two. Under this economic regime, cells could afford to grow larger and experiment with new ways of responding to different environments. That was the beginning of the evolutionary complexification that led to multicellular life, including us.

The descendants of the ancient collaborating bacteria still exist in our own cells under the name of mitochondria, and they have become our powerhouses. Each human cell contains from a few to thousands of mitochondria, and each mitochondrion contains a single bacterial chromosome that now encodes only thirty-seven genes. The other hundreds of bacterial genes having to do with mitochondrial function and regulation had long since migrated to the cell nucleus, where their replication is managed much more efficiently. The arrangement has remarkably resulted in new complexities, such as sexual reproduction as we know it today.

There is just one more phenomenon that belongs in this extravaganza of wonderment, and that is the origin of plants. Plants have mitochondria very similar to ours, and they also have another organelle known as the chloroplast. Like mitochondria, chloroplasts also have a residual bacterial chromosome and a unique fragment of DNA with which they contribute to their own metabolic story. Cells harbouring chloroplasts have a new option to harvest energy directly from the sun. This improbable collaboration involving a eukaryotic cell, and a bacterium also had impossibly long *a priori* odds and seems to have occurred only once.

Do these observations and musings matter in the real world? These days, astronomers are preparing themselves to look for signs of life on planets circling other stars. I anticipate that it won't be difficult to find life there at the level of complexity of our archaea and bacteria, but eukaryotes will likely not be easily found, if they exist there at all. The rules that derive from the universal Second Law of Thermodynamics don't seem to allow much room for deviation from our earthly pattern but, that may just be a manifestation of our failure of imagination. Perhaps nature on those planets has used its creativity to develop something that will be totally new to us.

LOOKING BACK THROUGH TIME'S TELESCOPE

In the second century, Claudius Ptolemy wrote *The Almagest*, the oldest work on astronomy still in existence. In it, he described a universe that was centred on the Earth around which all heavenly bodies orbited in perfect circles. The concept of circular orbits derived logically from Greek philosophy, with their emphasis on geometric perfection. The problem with Ptolemy's model was that it didn't quite fit all the available data, even in his time. There were two problems, the first of which was that of the wanderers: those five objects named Mercury, Venus, Mars, Jupiter and Saturn that traced erratic pathways through the sky, sometimes moving along the same direction as the stars and then turning to move in the other direction for a while. He named them "planets," the Greek word for wanderers, and seems not to have puzzled about them that much more. It remained for Nicolaus Copernicus who, fourteen centuries later, reconfigured the solar system to a model centred on the Sun and the planets in concentric orbits about it.

Ptolemy's second problem arose from measurements showing that the orbits were not perfectly circular. How could this be so in a universe modelled on divine perfection? His solution was ingenious in that he preserved the concept of circularity by proposing a system of epicycles. Perfect circularity was preserved through a very complex mathematical computation. Cumbersome as they were, no one challenged these calculations until the seventeenth century, when Johannes Kepler threw down his quill, so to speak, and said the Renaissance equivalent of "These are ellipses, damn it!" After that, the blinders fell from everyone's eyes, and the astronomers sheepishly responded with, "Why didn't I see that?" Thereby, a new window opened for modern observational astronomy.

My purpose in presenting this history is that Ptolemy was doing the very best he could with what he had at hand, and he is not to be faulted for having been wrong. It was not his fault that the age into which his work survived was blinkered and authoritarian. Over the ensuing centuries, the scholastics piled on an immense overburden of misconceptions that blurred the simpler facts of planetary movements, until the "Aha!" moments of Copernicus and Kepler. We also need to be open to the lessons that lurk for us in this story.

The last century has been one of ever-cascading success stories about our growing understanding of the universe. In my lifetime, we have learned that the Milky Way is not the whole universe; indeed, by the latest count, it is but one of perhaps two trillion similar galaxies within the visible universe, each with an average of a hundred billion stars. We have learned that most stars have planetary systems orbiting around them, and that there are billions of stars even in our Milky Way that might be able to support life. We were stunned to learn that the familiar physical matter of the universe accounts for less than 5% of the gravitational force, and that both dark matter and dark energy are yet to be understood. The discovery of the cosmic microwave background told us that the universe had a birthday, and the evidence from dark energy somberly tells us that it will also die. There is so much more, and all of it is amazing. Should we be preoccupied with drinking the ambrosia, or might there be more surprises in waiting still?

We are always more confident in our interpretations of scientific observations when observations agree with the predictions extending from our pre-existing hypotheses. In general, that has happened, but with one current glaring exception. The theoretical predictions and the measurements of a factor known as omega, the flatness index of the universe, are at odds by a factor of 10^{120}. However, measurements of the universe's curvature show that the accessible universe is flat within a very narrow margin of measurement. For the moment, measurement must trump theory, and I will wait to hear if that is in for revision.

MICHAEL FARADAY

Michael Faraday (1791–1867) was born into an impoverished family. When he was still quite a small child, his family, facing starvation, apprenticed him to a book binder who fed him, allotted him space to learn his trade, and gave him a place sleep at night under his workbench. His assigned task at the bindery was to sew together the fascicles prior to gluing on the backing and the attachment of the covers. He was said to have produced high quality work but received no formal education. At the age of twenty, he attended a series of science lectures by Sir Humphrey Davy and afterward, he presented Sir Humphrey with a bound set of carefully detailed notes he had taken at the lectures. Sir Humphrey was sufficiently impressed to hire young Michael to be his laboratory assistant. One of Michael's early assignments was as Sir Humphrey's valet on a European tour; a task that did little to cement the relationship. However, he seized the opportunities to engage European scientists with questions on a range of topics.

Step by step, and by dint of his intuitive attachment to scientific observation, he gained stature among scientists. One day, he noticed that moving a wire in the vicinity of a magnet caused an electric current to flow in the wire. There followed a long line of experiments during which he formulated the associations between polarity of magnets and directions of current flow that led up to designs for electricity generators and electrical motors.

Given his lack of a formal education, Faraday had no facility with mathematics. His modes of thought and the explanations he offered were more analogical than analytical. He was at his best when standing behind his now-famous desk, with the cut-out that facilitated his reach, demonstrating experiments at the Royal Institution. His demonstrations were said to have played to packed houses. It remained for his successor, James Clerk Maxwell, to develop the underlying mathematical theory for electromagnetism. Over the next two generations, that line of thought led to such abstruse concepts as relativity and quantum mechanics.

Faraday was widely recognized for his work; he was honoured with a doctorate from Oxford University and given the post of Director of the Royal Institution. He was granted membership in many international societies.

As his reputation grew, it became difficult for polite English society to continue to exclude him on grounds of his humble origins. The ladies, who controlled the dining and sitting rooms, continued to exclude him, but the gentlemen were determined to allow him in. The compromise they worked out was that on these occasions, Faraday would come in through a side door and be ushered into the kitchen, where he would eat with the servants and wait for the lords and ladies to finish their meal. Then the men would converse with him in the kitchen. One can imagine the spirited conversations about science and engineering that might have occurred over the piles of dirty dishes, carved-out bones, and empty wine bottles. It may have been such a situation that gave rise to a likely apocryphal anecdote in which a politician asked, "And pray tell me, Mr. Faraday, of what use do you imagine your invention [the turbine] to be?" Faraday prophetically responded, "I expect you politicians will find a way to tax it."

We now know that Faraday's invention of the turbine was an unwitting mimic of nature. The largest example of a turbine that we have is the Earth itself on account of the differential rotation of the solid and molten phases of the core. At the other extreme of size is the example of ATP synthase, an enzyme of which we harbour thousands of copies in every one of the hundred trillion cells of our bodies. In the Earth, it is the rotation of a conductor in a magnetic field that generates the electrons, and, in the enzyme, it is the energy of metabolism that creates the electron gradient that is essential for life. Lest you doubt the fulfillment of Faraday's prophesy, look at the line items of your next electrical utility bill.

At the outbreak of the Crimean War, the British Government asked Faraday to participate in the development of chemical weapons, but he refused the assignment on ethical grounds. He was a non-conforming Christian who believed he ought not to aggrandize or take steps to bolster his legacy. On those same grounds, he refused a knighthood and burial in Westminster Abbey. He died in 1867.

The world is better for having had Faraday in it. He would never have made it into a modern graduate school, as he had no undergraduate degrees, and he also would never have passed the entrance exams on account of his incorrigible ineptitude with mathematics. I don't believe that anyone ever intended that the path to a scientific career should be obstructed for exceptional people. There have been other people without formal education, such as Milton Humason, who began as a mule driver transporting Edwin Hubble's telescope parts to a mountain top, and who stayed on to become, *in seriatim,* the observatory's janitor, then a research assistant, and finally, an investigator who collaborated on Hubble's estimation of the expansion rate of the universe. I hope there are more such people around, and that they can still find their way to the forefront they merit.

ODE TO READING

On the first day of summer holidays, when I was eight years old, I realized, with a sense of panic, that I had brought nothing home from school to read over the next two months. The only books we had in the house at that time were Dad's German Bible and another tome that was filled with horrifying pictures of the second coming of Christ and the torments of the damned, which had been dropped off by an itinerant missionary. After reflection, I decided to fill the void by writing my own stories. I had a notebook and pencil left from school, and these were resources enough. The notebook soon filled up with a pointless story, and that was the end of that. The notebook drifted down to the bottom of the sock drawer, and I never thought about it again. That is, until it resurfaced more than forty years later when Mother mentioned having found it and having been disappointed by my portrayal of her. What did I know about the world outside of our poverty-stricken rustic prairie farm!

I've been a fanatic reader since the afternoon of my first day in school, when the symbols on the page of my first book began abruptly to make sense. Thereafter, it never occurred to me that my capacity to read was in any way limited by considerations of vocabulary, content or age appropriateness. I worked through large sections of the weekly *Free Press Prairie Farmer* newspaper, including the serialized novels that were meant to relieve the otherwise unbroken tedium of prairie life in the pre-TV (and, for some homes, pre-radio) era. In retrospect, these novels were the essence of pulp fiction and contributed little to character building. Had Mother known what I was reading, she would have had it in her wood-burning kitchen stove in a moment. I recall one plot that featured a doctor being approached by panicked men galloping into town on sweaty plow horses and shouting, "Doc, ya gotta come now. Bessie's time has come!" Whereupon, the doctor leapt onto his own horse, but not before grabbing his hat. There was something gallant about galloping across the countryside on a mysterious errand and with a red rubber tube coiled up inside the hat band. I told Mother that I had decided to become a doctor. To this, she said, "You can't get your nose out of a book for long enough to do your chores; that's no way to be a doctor."

Early in tenth grade in Kindersley, I discovered the town's library at the back of the firehall. It was open on Tuesday evenings and run by an elderly lady who seemed to be mostly alone. She welcomed my visits and allowed me to take out up to three books each week. I adopted a very simple way of choosing books, the next three across each shelf as I came to them. Novels, anthologies, history—nothing was safe from my curiosity. I used the ten-minute breaks between classes at school to complete homework assignments so that homework wouldn't interfere with my reading during evening study periods at the high school residence.

To this day, I am still an omnivorous reader, but the books I enjoy most are typically not on bestseller lists, and they are often not stocked by local libraries. Ergo, I buy the books I read. The rooms of my house are, as in the dreams of my youth, lined by bookshelves. For several reasons, I've not been fully satisfied by the emerging experience of ebooks. Once I signed up to an ebook service, but my email immediately filled up with unwanted promotions for books that fell far outside of my genre preferences. More substantively, ebooks miss the sensuality that emanates from a well-bound book; it's not unlike a glass of good wine, where the glass contributes to the appreciation of the aroma as well as its feel in the mouth. There is so much more to a book than just its linear verbal feed. The heft of the volume, as well as the quality of the binding, are essential elements of the experience. Ebooks are more like wine from a paper cup.

For most of my life, I imagined that my personal library would one day be appreciated by those who would come after me, but I've come to realize that people do not read books anymore At least, they do not read them the way I do, stopping to savour the antiquities of spelling or unique turns of phrase in a tale set in a far-off place. I am always aware of the possibility that I may change my opinion on a topic because of what I read. When I am done, I reverently replace the book on the shelf with the intent of one day re-reading it. My family, however, have on-line access to all of this, and a bookshelf filled with the finest distillations of the historical human mind is no more inspiring to them than a row of empty wine bottles in a basement TV room. Perhaps the best hope for my library is that it will eventually be reduced to its value as recycled paper. But how is the joy I had in learning its contents and accumulating all these trophies to be honoured? Ashes to ashes . . .

ON CHANGING ONE'S MIND

It is an advantage of a long life that one is afforded more opportunities for a change of mind, which allows it to iterate toward what should be progressively better solutions of every kind. It is strange that, despite the number of times one gets to do so, it doesn't become easier. One carefully studied illustration of this generic difficulty is found in Daniel Kahneman's book, *Thinking Fast and Slow*, in which he recounts his observations of stockbrokers who, despite experience, routinely hang onto declining investments for far too long, always with the expectation that the values will turn around. As Kahneman's data show, experience in this exercise does not improve performance. The problem lies in the fact that most people want to earn money on the market quickly, and only a few people care to think deeply about societally appropriate investment objectives.

In my experience, changing an opinion is always difficult, and it requires hard work and attention to details. On the first day of my education in obstetrics, I learned that despite the fact that abortions were then defined under the criminal code, complications of this intervention were the leading cause of death among women of child-bearing age in Canada. At first, I assumed the solution would be to enforce the law more tightly. Despite accumulating evidence of harm from criminally induced abortions, it took me about fifteen years to realize that the problem lay not in the evil of abortion but in the desperation of people facing undesired pregnancies, and definitely not in evil women. It gradually dawned on me that the solution might lie in the direction of asking them what they most needed. When that actually happened, the gangland abortionists became socially irrelevant and disappeared. Looking back, I realize now that what it took to change my opinion was an amendment to my understanding of the nature of justice; it is less about punishment for mistakes and more about healing for people who hurt in any arena of life. After that, my opinion of abortion became a mere corollary to an amended understanding.

A seminal event along that journey occurred when I, thinking that there would be space allotted for critical exploration of issues, briefly joined a right-to-life organization. It took me only a short time to learn that the group held ideology that trumped reason at every turn, and there was no room for probing discourse with them. The experience may have been the jolt I needed to move my subconscious cognitive dissonance to a conscious level. I've now had sixty more years of adult life to consider and reconsider the rights and wrongs of abortion practices and the implications of related social policies. I now realize that an ethically driven direction does not always lead to a perfect outcome but should always help to minimize the pain of others.

I've used this experience to reflect on what it may take to change an entrenched opinion, mine or another's, on any subject. I could also illustrate my point with anecdotes from my political and philosophical evolution. My first set of ethical values and moral principles were developed for me in childhood, in the framework of evangelical Christianity, where questions of right and wrong in society were directly equated with good and evil in the religious sense. Given those references, there was no possibility that any reasoned argument, however intellectually powerful, could break through my opinions; my entire framework of values had to crumble and be rebuilt before reasoned exploration was possible. Given a new set of reference values, which happened in my mid-forties, questions about abortion became mere codicils appended to a new main game.

This works both ways. Every week, there is a pathetic parade of anti-abortionist campaigners lined up in front of my former hospital. I say "pathetic" because they have been focused for forty years on a single activity instead of on addressing a possible value-system revision. I'm sure they have been confronted on the street on many occasions, but they are still there. If only they would kick up the debate to the level of their value framework, then there could be productive discussion. In the process, we would also resolve many other ethical issues. That change would require an inner debate of each person with their own selves—not only with others—and that is the part of changing an opinion that is so hard to do.

So, what does the poor stockbroker have to do with all this? Possibly everything. For so long as she is solely focused on the primacy of immediate profits to report to the next shareholders' meeting, there can be no systematic healthy investing and no possibility of using investments for any purpose except to slake the appetite for greed. What ethical investing should be about is not solely profit-directed, but also with an eye on long-term sustainability through support of human dignity and careful cherishing of our limited resources.

OF BOMBS, PLOWSHARES, AND MEDICINES

When the first atom bomb exploded at Alamogordo in the spring of 1945, Robert Oppenheimer, the scientific director of the program, leaned his head against the wall of the bomb shelter, as in mourning, and was reported in one account to have said under his breath, "Now we are all bastards." According to other accounts, he quoted from the Bhagavad Gita but to the same meaning. What he meant was that with this weapon, the Allies had given up the moral high ground, the standpoint from which only the enemy slaughtered innocent non-combatants in terrifying and heartless genocides. The Allies were supposed to have saved humanity from the despotisms of Germany, Italy, and Japan. Reciprocal genocide was not supposed to be a card we played.

But then, on August 6th and 9th, a quarter-million Japanese civilians were immolated in two bright flashes of light. The justification offered for the bombing of Hiroshima and Nagasaki was that it brought about the Japanese surrender and spared the lives of many thousands of American soldiers, lives that would otherwise have been lost in the invasion of Japan. However, as Gar Alperovitz outlines in *The Decision to Use the Atomic Bomb* (1996), this was a false excuse. The Japanese had, in fact, been petitioning the Allies for terms of surrender for a month prior, and the terms they offered from the outset were essentially those that the Americans ultimately accepted. Why was their plea for surrender ignored for so long?

It became clear over time that President Truman's strategy in deploying the bombs was his intent to alarm Joseph Stalin who, he correctly anticipated, would be a difficult ongoing Allied partner after war's end. However, the surprise fell flat, because Stalin had a very competent spy, Klaus Fuchs, who was ensconced in the Manhattan Project from 1943 onward, and who wasn't discovered until 1948. Thus, in addition to the senseless nature of the attacks, the victims of those two bombs lost their lives to no purpose. If we only had resources enough to right the wrongs of the past, it would, in my opinion, be justifiable to try Harry Truman on war crime charges.

The Manhattan Project, which culminated in the construction of the atom bombs, was managed by General Groves, who instituted a strict need-to-know security system, such that only a few people actually knew what the goal of the project was. Individual workers and teams knew only the parameters of their own assignments and not the entire mission. When the nature of the project was finally known more widely, a lot of scientists and technicians suffered severe moral agitation for their unwitting complicity, and there was widespread concern that something had to be done urgently to bring clear evidence of benefits from nuclear science to society, and thereby to somewhat redeem it from the stain of being a mass murder weapon.

Broad areas were selected for urgent development to this end, and the two that topped the list were nuclear power and medicine. There were members of the public who, at the time, were suspicious of these initiatives as distractions from the main agenda of building a nuclear arsenal as well as a fleet of nuclear submarines and aircraft carriers. To some extent, they were right about that. However, the benefits of nuclear science to society since then have been real and enduring, and they have extended into innumerable additional areas.

I am intrigued that we had such an unprecedented surge of new knowledge after WWII, and that we have now, perhaps inadvertently, attained such a high degree of economic inter-relatedness among nations that war is not so often a favoured means of resolving international tensions.

Some of the important influences in this direction, such as the development of CERN (European Centre for Nuclear Research), were not primarily rooted in governmental decisions at the time. CERN, to this day, is a scientific collaboration that is devoted to improving the understanding of matter and shares new knowledge among members without payment of royalties or other fees, except it will not conduct research of primary benefit to any military sector. More recently, a similar consortium of Middle Eastern countries (Turkey, Israel, Jordan, Iran, Palestine and Egypt: could there be a collection of more unlikely bedfellows?), known as SESAME, has begun to function in Amman along the model of CERN and similarly promises, through sharing of intellectual discoveries, to provide shared opportunities for economic development in the region. There may be alternatives to confrontation after all, but it's not likely to be easy. The image of wildflowers blooming on the graves of those who were lost so many years ago comes to mind.

ON GETTING A CHARGE FROM LIFE

It is sometimes the case in science that the research community comes up against a seemingly impenetrable obstacle that holds off all assaults for a long while until someone thinks to restate the problem, after which a quantum leap in understanding follows. This may turn out to soon be the case with the problems posed by the origin of life.

There are living things in millennia-old Antarctic ice cores, as well as along the hot crater edges of undersea volcanoes that thrive at temperatures above 120°C; others exist at the bottom of the deepest newly drilled oil and gas wells. Life in its familiar forms flourishes where there is oxygen, but the abundant biomass in the rocks of the deep earth is hundreds of times greater than what exists on the surface, and it gets along nicely without it. Indeed, oxygen is toxic to these life forms in the deep rocks that depend for their metabolic energy on electrons harvested from crystal surfaces. That system of life runs slowly and may require millennia to replicate just once, whereas one that uses oxygen will do the same in fifteen minutes. We've learned about this deep-dwelling and penetrant life only in the last few decades.

It comes as a surprise that all of life uses the same core metabolic pathway; eight linked reactions taught to students as the citric acid cycle. This shared bit of biochemistry has been referred to as a biochemical fossil, but it still operates in our own bodies. Aerobes are the life forms that use oxygen (eukaryotes and some bacteria), and they run the cycle in the direction of oxidation, releasing their "used" electrons as carbon dioxide, whereas anaerobes (ie, autotrophs) run it in the other direction, and electrons are delivered to carbon as methane. In each case, electrons at high potential energy give off energy to drive the reactions of life.

There is a new question forming about the origin of life: in what way was the Earth out of balance such that life became the easiest solution to resolve its stress? I owe this sensible formulation to Professor Eric Smith, who suggests that Earth is in reality a battery that is kept charged from three energy sources: we all know these as the magnetosphere, radioactive decay, and sunshine. In short, life is the way that electrons go to ground most easily. Is this in any way a solution to the question of life's origins? I doubt it, but the proposition does allow for a measure of disciplined hypothesis generation, and possibly some experimentation. The stuff one finds in nature's attic is amazing.

ON PLAYING GRACEFULLY

The actor, writer, and sometimes progressive advocate, Stephen Fry—in his sometimes Churchillian tones—bemoans the loss of people from our society who take joy in "playing gracefully with ideas." So, do I. There is reason for the loss; it takes time to incubate thoughts worth sharing, and the process can be lonely. It's so much easier to extol a sports team's performance to enliven a convivial drink, and if you spoke instead of a mere idea from another realm in the same voice, you might not be invited back, especially if someone else around the table voiced an opposing opinion. As a result, and because most of us haven't learned how to deflect dissent into the arena of reasoned debate, a lot of really important conversational potential is made off limits in polite society. Fights are ugly, but debates can be beautiful and elucidating; moreover, the participants must master the finesse needed to negotiate their way along the tensions existing between the extremes. It bears repeating; "play gracefully with ideas."

I once watched a debate among a group of astronomers that included the chief astronomer at the Vatican Observatory, as well as some who were of militantly atheistic views. In the group was Frank Wilczek, now in his seventies, who received a Nobel Prize in Physics for work he did before the age of twenty-one. Frank seems to be an amiable person who is always smiling and typically says little, despite his prestige. The referenced debate was spirited, and in the excitement, the participants were beginning to speak over each other when Frank intervened with just a few sentences in his softspoken tones and interjected a fresh line of thought that immediately cooled the emotions and redirected the debate in a more positive direction. This is a man with whom I would like to have dinner, if only for the opportunity to learn a little more about the art of conducting meaningful sustained conversation.

Clifford V. Johnson is a physicist who has taught himself the skills necessary to write about science topics in graphic style (i.e., comic books). In his style, allusion to the gravitationally induced distortions of space-time are opportunities to begin, and not to end, a rich dialogue about such profound facets of the real world, whether they arise on the bus or in a coffee shop or in the elevator.

There is nothing I would like more than to be able to open all our taboo topics to honest and unguarded exploration. Why does almost every participant feel driven to take sides in a politically or religiously tinctured debate? There might just be neutral evidence brought to bear that might enrich our further reflections. There might just be a relevant exculpatory historical vignette to offer to the discussion. Or, at the end, there might be only a sticky residue of indignation, or yet a benediction of awe and wonder. Whatever the outcome, we should still allow ourselves to think that we had played gracefully and strengthened our friendships in the process. Science communications demand no less.

ON FINDING THE LOWEST DENOMINATOR

I remember so well the occasion on which I was introduced to the World Wide Web. It was June 1994 on the day that fragments of the Shoemaker-Levy comet were smashing onto the surface of Jupiter, each releasing more energy than hundreds of nuclear bombs. Beyond the awe induced by the spectacle itself was the speechless amazement with which I viewed this new enabling technology that permitted its real-time viewing on an office computer.

There was, in hindsight, a cumbersome File Transfer Protocol (FTP) required to access the early web before there were browsers, but the ease of visiting libraries around the world from my office desk made writing and revision of manuscripts so much easier. No longer was it necessary to book myself out of the office for a couple of hours for a drive across the city to retrieve a handful of papers from the library to complete one of my own. Of particular interest to me, then, was also the new opportunity offered to readers to make online critical commentaries on public events and other publications. Every member of the public could now aspire to be an oracle on the standard of Cicero and capable of incisive comment. Here was an opportunity for everyone to contribute to accountability of the whole of our society. Who would not be seized by the opportunity to practice and become an orator functioning at the highest level of competence?

So, what actually happened? To say that I am disappointed is to understate the case. It has become apparent that most of humanity is more deeply impacted by emotionally conditioned impulses than by opportunities to engage in the hard work of thinking. The ability to hurl invective from behind the barrier of prejudicial anonymity seems, for most commentators, to be too much to resist. The core of an argument contained in a news item is invariably lost in favour of sublimated pugilistic brutality, and the occasional brave soul who tries to keep the remarks on topic is destroyed.

Unfortunately, the case with many science-directed sites is no better; a few days ago, I listened to an extended discussion of the weighty subject of gravitational waves, and afterward scrolled to the comments only to find it taken over by issues of the brevity of some skirts and hairiness of some legs. Oh yes, I forgot to say that the presenters of this leading research were all women with recognized accomplishments in theoretical physics. Today, I took in a YouTube discussion concerning cognitive psychology, and the ever-ready commentators were quick to dismiss a thoughtful discussion of the psychological implications of swearing to denigrate the scholar's Jewish name. You can also count on Biblical literalist trolls to be on the site like a swarm of hornets when the subject is related to evolutionary biology, and their invective is no less toxic.

For this essay, I'm not so much concerned with the ongoing degradation of manners and respect as I am with the fact that these abuses of the Internet are fueling the raging fires of a dangerous populism according to which one does not need to be an expert on anything to express an opinion on everything.

I use a disease-based model of the spread of epidemics to understand the transmission of our degraded behavioral norms and find many parallels. Fascism, like the flu, is back. The Internet has become the major mode of transmission, and we are desperately in need of a new analogue of a vaccine if major conflicts are to be avoided. Abruptly, experts who were once esteemed for their valued nuanced opinions are now derided as *elites* deserving only abusive derision and dismissal. When did a total lack of prior experience for one of the most difficult jobs in the world become a desired characteristic of the recent holder of the highest office in America?

There are too many broken things in this scenario to suggest that any one repair would do for all. To do so would be presumptuous in the extreme. I cannot be sure whether current abusive Internet behaviours are a cause of disease, or only a symptom. I've given up on being heard as a commentator on news sites because nothing I can say will make a difference in the prevailing cacophony. I have taken to talking to myself in the most elevated tones I can muster on the off chance that something I mutter in my private despair will be overheard and somewhere activate a butterfly effect.

ON SEEING STARS

Some years ago, I had an accident in which I lost consciousness, then spontaneously came back to myself, only to lose it again some days later. I finally reawakened after an emergency operation to drain a bleed (subdural hematoma) from my brain. At no time throughout that experience did I see any stars! When I awoke, my mind was clear, and I was able to complete reading Lee Smolin's treatise about the nature of time. But walking was difficult, stairs were initially impossible, and I couldn't button my shirt. Clearly, my recovery period was going to (and did) take some time. My immediate problem was how to occupy my mind for as long as this was going to take.

I was pleased to find the Kepler Project, which at that time, was looking for citizen scientists to review data regarding the newly discovered planets in extrasolar systems. The Kepler space telescope first began to make observations in the form of photometry readings on 150,000 stars at thirty-minute intervals in 2009, and it kept that up for more than five years. The resulting light curves showed a transient dimming of the light each time a planet passed in the telescope's line of sight.

There then followed a protocol of verification with Earth-based telescopes to rule out false positive readings as from interfering dust clouds in space. Ultimately, more than 5,000 exoplanets were verified. Since the test is positive only if the planetary plane of rotation is in the telescope's line of sight, one can calculate that there are many times more planets in that field of view than were actually visually confirmed. By extrapolation, there may be more than one hundred billion planets circling low energy stars resembling our Sun within the galaxy. During my rehab time, I monitored stars via the Kepler data, and I observed quite a lot of interesting phenomena, including some of those candidate planets.

What we have learned from the Keplar observations is that the formation of planets is integral to the processes that form their stars from the same dusty Hydrogen/Helium clouds. Initially, lazy clouds of gases, and the debris of previous super novae created by the deaths of earlier generations of stars drift more or less undirected through space. Over time, the action of gravity focuses matter on the mass of the future star as it becomes compacted and takes on a rotation that is the sum of all the vectors that operated in the previously amorphous cloud.

Not unexpectedly, the first exoplanets after the Big Bang turned out to be very large; the so-called Super-Jupiters and gas giants composed mainly of hydrogen. Their size made them easy to detect. Then, as techniques improved, smaller planets, super Earths and Earth-like, also turned up. In some cases, multiple planets around a single star, like our solar system, were detected.

Of course, the burning question that pops out of these observations is whether there is anyone on any of those planets who is looking back at us. What are the odds that there is life over there? Granted, many planets are too close to their stars (which makes them too hot), or they are too large (i.e., they have too much gravitational attraction), and some may have no liquid water. Still, one is left with a sampling of candidates that seem to be in the range of habitability and that merit further study. The Magellan Telescope, now under construction on the Atacama Desert plateau in the Chilean Andes, and the 30 Metre telescope under construction in Hawaii are designed to identify features that will inform us about those planetary atmospheres. Later in this decade, the VLT (Very Large Telescope) will also be completed.

Do the math. If there are one hundred billion Earth-like planets circling stars in the Milky Way, and if only one in a thousand supports, or has ever supported, life, then there are about one hundred million life-bearing planets in the galaxy. From the Hubble Telescope observations, there are an additional two trillion galaxies in the visible universe with similar odds for each. Someone ought to be writing the majestic theme music by which to herald the discoveries that are yet to come. Or did I bump my head once too often?

PARABLES FROM A SHOELACE

The anatomy of a shoelace is relatively simple. It consists of a length of cord that terminates in a bit of celluloid known, especially to crossword puzzlers, as aglets, and whose function is to ease the task of threading the lace. We all know how frustrating it is to deal with a frayed lace. It reminds me of a chromosome, one of our twenty-three pairs of doubled helical strands of DNA whose ends terminate in an analogical structure known as a telomere. Like aglets, telomeres become more fragile with age. Each time the cell divides, its telomeres become shorter, and after they disappear completely, the cell cannot divide again. There is ongoing research to identify the possibility of therapies to restore the telomeres to their youthful selves and, correspondingly, extend our life expectancy. Would I want that to happen to me? An old, scuffed up shoe with a new lace is somewhat incongruous, and I've never noted that combination to mark a fashion trend. Do I, at my present age, want to be recertified for another lifetime? Will society tolerate a new population of resource-draining super-seniors? I'm not sure that telomere restoration is the way to go except as the ultimate in self-indulgence.

In a pensive moment, I sense that neither the shoelace nor I have a shot at immortality. Clearly, we were both engineered with an expiry date in mind.

The images of aging telomeres and broken aglets stand for whatever they are. Yet there is another comparison of shoelaces and chromosomes to be explored. Notice that each pair of our twenty-three chromosomes has a constriction somewhere along its length where the two strands come together. This region is called the centromere, the point from which the chromosomes are pulled apart when the cell divides to make two daughter cells. There is one centromere per chromosome.

Now recall a situation in your past when you were lacing up a tall boot and the lace tore, leaving you with no option but to tie together a pair of shorter laces from your running shoes to get on with the urgencies of your day. This gave you a lace with a pair of aglets on the inside near the knot, as well as the other pair at the outer ends. Something like this has happened to the human chromosome number two. All other apes have twenty-four pairs of chromosomes. Why do we have one fewer? Close study has revealed that our chromosome number two has a second centromere that no longer functions in cell division, and there is an unusual pair of telomeres buried in the central region of that chromosome. Somewhere in our past (circa six million years back), chromosome two of an ancestor fused with another chromosome to achieve our reduced complement of only twenty-three.

There are other circumstances in which chromosomal fragments have fused to produce new arrangements. The normal manifestation occurs in generation of germ cells in which a process, known as recombination, rearranges genes from parental germ cells. There are also some rearrangements that can be catastrophic, as when a cell has been exposed to ionizing radiation and chromosomes randomly broken, then repaired incorrectly with typical results of cell death or cancer following.

The state of our chromosome two is very strong evidence at the molecular level that evolution was at work, using what it had available to work with to craft the conditions leading to us.

OUR DEBT TO PAUL EHRLICH

In 2008 I was surprised by an invitation to a conference in Nuremberg to commemorate the centenary of the award of the Nobel Prize to Paul Ehrlich. Ehrlich was a powerhouse academic physician who contributed in a durable way to three areas of medicine, namely: immunology, targeted drug therapies, and oncology. The prize was awarded for his achievements in immunology, which led directly to routine childhood immunizations against diphtheria and whooping cough. Later on, he successfully treated syphilis, which was then the most common cause of disease and disability in the Western world. On the oncology front, it turned out that he was reaching too far into the future for advances that would not be made until several additional generations of additional work had been done. He was also a legendary chain cigar smoker who went through thirty Cuban cigars daily, which was a likely contributor to his early death from a heart attack.

The conference I attended had clearly been organized by an informal group of academics who intended to use the occasion to not only honour the memory of Ehrlich, but also to have a fun weekend with a supposedly small number of colleagues in an intimate setting. To everyone's amazement, two thousand people showed up to register at the business school, which had rented out only three classrooms for the weekend. As I entered the first session, I was handed a copy of the program, and noticed to my amazement that I was, without advance notice, the designated chair. I had no time to pre-read the abstracts or prepare questions, much less to think on the pronunciation of presenters' names, but it was fun.

Ehrlich's life work was guided by a hypothesis that he formulated while still a student. It was then being noted that animal and bacterial cells could be stained by various dyes to reveal distinct components of cells, and that there were characteristic changes in specific disease states that made the stains diagnostically useful. He reasoned by extension that these distinct chemical distributions could be extended to include therapeutic chemicals. This was the birth of the concept of the "silver bullet," later known as disease-specific therapy. Keep in mind that to that time, there were no curative therapies for any diseases, and many of the world's leading physicians were sceptical of his likelihood of success.

In the realm of acute bacterial diseases, such as whooping cough and tetanus, the silver bullet proved to be immunizations that stimulated the production of disease-specific antibodies. However, there were many other infectious diseases for which the treatments were not so simply achieved. Two of the most difficult public scourges of the 19th and early 20th centuries were tuberculosis and syphilis. At that time, it was not yet widely agreed that tuberculosis was infectious in origin. The spirochetes, however, fulfilled all Koch's requirements of an infective agent and were recognized as the infective cause of syphilis.

The medication that eventually proved to be curative for syphilis was named Salvarsan 606. The name is significant and indicative of Ehrlich's persistence in trials with 605 previously failed agents. Also significant is the word "trials." The entire historical pharmacopeia supporting medical interventions up to that time were justified through the authority of physicians without obligatory, objective evidence of efficacy. Beyond the immediate sphere of syphilis, Ehrlich, almost coincidentally invented the clinical trial as the basic element of future therapeutic advances. I cannot be clear in my own mind that he himself fully appreciated the significance of that accomplishment. His methods were primitive, lacking critical elements such as peer review and prior animal toxicity studies; he came to suffer for that when he was dragged into court because patients had died in some of his trials. However, we must remember that other elements that figure largely in our understanding of modern clinical trials, such as bureaucratic regulation, independent ethical reviews, and informed consent, were not articulated until long after his death.

Ehrlich was severely criticized by some elements of "proper" society, especially by the churches, for having chosen syphilis as the disease most urgently needing treatment. They argued that the prospect of cure would break down the floodgates against immorality, all the while failing to notice that the known consequences of venereal disease had never before restrained sexual behaviours.

I am grateful for the privilege of attending the centennial celebration of Ehrlich's award. My own contribution to the meeting was in the form of a discussion of radionuclide therapies as one form of those once-imaginary silver bullets. Today, the thrust of Ehrlich's central hypothesis has been validated countless times, and we are indebted to the man who persevered despite having logged 605 successive failures before he found the first silver bullet. And we take some inspiration from him as an icon of the new field of theragnostics.

LOOKING THROUGH THE
FAMILY PHOTOS

The popular conception of human evolution is heavily conditioned by the images brought to us by the likes of *National Geographic*. I'm alluding to a double-fold that portrays a hairy, small-brained chimp groping through a bush on the left side of the page, followed by serial silhouettes that gradually morph into an upright, Olympian human body, athletic, hairless, and racing victoriously into the Sun. Implicit in this montage is that human evolution has been purposefully directed toward a goal; ie, us in our present form. Only, it wasn't like that at all. Evolution was an undirected process that could have taken—and did take—many different probing courses and followed into some blind ends, as evidenced by the many twigs on the evolutionary bush, most of which died off and were abandoned a long while ago.

Depending on how the data are grouped, there have been somewhere between a dozen or as many as twenty human species in history, all of which originated in Africa and related to each other through a common ancestor that we share with the chimpanzees about six million years ago. Our branch of the bush, *Homo sapiens*, is the only one left with life on it. It may be instructive for us to consider how the others were similar to or different from us, and to ponder why they lost the contest to race into the *National Geographic* Sun.

Within the last decade, we have learned how to extract DNA fragments from fossilized bones up to nearly a million years old, provided they were found in dry and cool conditions. It needs to be said that human(oid) history goes back much farther than that, but the DNA sequencing tools won't take us further back. It seems that the hydroxyapatite crystals in bone act to stabilize DNA over millennia, and we do have complete sequences for modern Humans, 600,000-year-old Neanderthals, Denisovans, and modern chimpanzees.

Point mutations occur at a frequency of about one/million base pairs, or about thirty/generation in each replication of DNA and from that we can construct a clock to measure how long ago the species separated from each other. Having done that, we can say that humans became distinct from chimpanzees about six million years ago. The original Neanderthal migration into Europe and the Middle East, perhaps five to six hundred thousand years ago, provided a geographical separation, which allowed for related groups that initially remained in Africa to undergo a period of separate evolutionary development. Evidence indicates that Neanderthals and homo sapiens found each other again about 60,000 years ago, when the latter migrated into Europe, and that they interbred extensively at that time.

We learned more recently of another branch: the Denisovans in southern Siberia. All we have from them is a complete genome that was determined from a single small finger bone fragment (the distal epiphysis of the intermediate fifth phalanx) and one other fossil that was the progeny of a Neanderthal female and a Denisovan male. There is also evidence of Denisovan DNA in the modern Tibetan population by way of a hemoglobin mutation that facilitates oxygen transport at high altitudes.

Most of the older African fossils have not yielded DNA because the conditions in tropical climates did not preserve DNA. However, the techniques of classical archeology still do teach us a lot. Some of the partial skeletal remains from the Great Rift Valley in East Africa in one species demonstrated feet with opposable great toes that would have facilitated tree-climbing. Other, later specimens from the Rising Star Cave in South Africa include a species (the *Homo naledi*) with a fully human foot, but with persisting features in the hands and shoulders consistent with retained ability to climb trees. A common finding in all early humanoids is a small brain; the *Homo naledi* brain was only slightly larger than an orange. The appearance of larger brains over a relatively short time presents an intriguing challenge, because the larger infant head required concurrent development of a female pelvis capable to match fetal head size. I wonder whether the move from trees to the ground and the altered weight-bearing factored into pelvic changes that effectively allowed for the selection of larger heads.

The *Homo naledi* were discovered in 2012, have been dated by several isotopic clocks to between a quarter and a third of a million years ago. They disposed of their dead in a remote spot at the back of their cave. Recent observations of cave wall markings suggest that they attempted to make art. If confirmed, this would be the first art made by a non-human entity. We will probably never know why they did what they did.

I'm impressed by the development of exquisite technology in archeology to elucidate small details of the past—for instance, the analysis of dental plaque to extract clues about diet. The ratios of carbon-12 to carbon-13 in plaque or in fossilized feces can tell us whether they were eating food from the sea in any amount. I'm also impressed by the sensitivity and protection of small details of modern archeological digs: at the Rising Star cave site in South Africa several sets of the three small bones from the middle ear were recovered intact.

Little by little, we are learning about the very large, the very small, and the very old in spheres as different as the origin of the universe, the beginning of life, and the origins of human culture.

Nevertheless, many questions remain open. For whose benefit did the Neanderthals make their cave paintings more than 50,000 years ago? Is there a message for us? Are we reading the messages correctly? Did the *Homo naledi* or any branches of *Homo erectus* leave something for us? I imagine their messages are all saying variations of, "You must taste the apples in the garden out back." If we all share a single characteristic with our earlier cousins, it must be that of curiosity. For all the trouble that curiosity has gotten us into, both in history and in mythology, it makes life exciting.

MEDITATIONS AT THE ATM

Lilianne was always a meticulous planner; no anticipated adventure was ever allowed to retain even one element of reducible uncertainty if it could be foreseen. So, when we visited the Large Hadron Collider (LHC), then under renovation, at the CERN laboratories in Geneva, she thought we ought to travel with a supply of Swiss currency, "just in case." To this, I replied that we were in the era of the credit card, and that I hadn't encountered an occasion in the past decade in which cash had served a uniquely useful purpose. She relented, and we travelled.

At the Geneva airport we were loading our luggage into a taxi when I asked the cabbie about payment terms and was informed that he only accepted Swiss Francs—no cards, no Euros, and no American dollars. After we unloaded the cab and found our way back to the airport bank, its ATM declined my credit card. It was only after Lilianne produced her card that our problem was resolved and, my humiliation complete, we were able to proceed as planned. In her true elegant form, she never rubbed it in.

I don't recall when it came to me that currency transactions are a good preliminary analogue to Albert Einstein's famous equation for Special Relativity.

1. Swiss Francs = Canadian Dollars x the exchange rate.
2. Energy = Mass x exchange rate (in this case, speed of light squared).

One could say that in this sense, the LHC is an ATM that accepts matter (protons) and, after accelerating them to relativistic velocities, delivers a pulse of energy. When two such pulses collide, the detectors allow us to see what one can purchase with these withdrawals—in this case, a momentary glimpse into the subatomic particle shower that characterized the instant of matter creation in the Big Bang.

Now I have to come clean and admit that I haven't yet told you the whole story. When Einstein was deriving his famous equation, he was driven by mathematical considerations to tag on a factor, lambda, that he thought necessary to stabilize his model of the universe. However, many scientists of the day insisted that the factor was unnecessary because the universe was then thought to be eternal and unchanging. In response to the pressure of their collective opinions, he took it out. He regretted it later when it became apparent that the universe had had a birthday and that it is continuing to evolve and expand. A universe without Lambda would be unstable in the manner of a pencil balanced on its tip. Einstein spoke of the moment he gave in to peer pressure as "the biggest mistake of my life." The fact of the universe's expansion (in short, the lambda factor) predicts that it will eventually die and dissipate into nothingness, but that is the price of stability in the nearer term.

Back to the ATM, the equation I gave above for currency exchange is also incomplete; it needs its own lambda factor, better known as the service charge. You may resent their high cost, but would you trust a banking system that consistently lost money? At the same time, you know that banks, though profitable, are mortal and may one day disappear in response to new innovations such as e-cash. We also have intimations that the end of the universe is waiting for its moment. Humbling as it may be, the next time you are waiting for your cash to slip through the slot at an ATM, let it be an existential moment.

REFLECTIONS ON DREAMS

For the ancients, dreams were loaded with portents and omens of forthcoming apocalypses and/or jubilees, the understanding of which often required interpretation by a local wizard, possibly abetted by other rituals such as a reading of the entrails. With such information in hand, emperors made portentous decisions that may have altered the course of history, including whether to start a war or marry off their children. In modern times, we drifted off from experiencing dreams as life-altering in this manner, instead seeing them as somewhat instructive about our inner aspirations, egos, and frustrations. Since the end of the Freudian era, we've also seen them as indicative of perhaps nothing more than a bad conscience, a disordered digestion, or an onsetting chill after the blankets slipped over the side of the bed in the night.

Dreams are such a mixture of things—a medley of twisted reality blended with something between the imaginary and the confabulated, and everything topped with traces of seemingly unconnected mystery. "Where did *that* come from?" is a common reflection on first squinting into the glaring morning light in the bathroom mirror.

In modern medicine, the information that physicians need to direct care comes to us mainly on screens and printouts; our dreams might, unsurprisingly, feature clinical scenarios drawn from our days. Throughout my working life, I had recurrent dreams that seemed to arise in hospital settings. They had a common theme in that at some point, I needed written information in order to advance the action in the dream. I never could read any materials that appeared, and the effort to do so invariably resulted in my waking up. Why, I ask, can my dreaming self not read these purported documents? Why, in my dreams, did I never have an opportunity to synthesize the data and achieve a brilliant life-saving diagnosis? Is there such an entity as dream dyslexia? I'm guessing that the dreaming and reading functions of the brain are inaccessible to each other. Perhaps, the dreaming brain is unable to bring the necessary level of intellectual discipline to bear such that information on the page would make even vague sense. Perhaps my dreaming self is so unimaginative as to be unable to confabulate to the extent required to fill the gaps.

These reflections have me wondering about historical figures who claimed to have received inspiration and direction through dreams. Was Joseph being prudent when he counselled Pharaoh to save for a rainy day, or was he merely lucky when he foretold of an oncoming famine? Was Saint Paul's dream of the invitation to evangelize Macedonia real, or only a *post hoc* confabulation to justify a decision already taken? How, in nineteenth-century America, did Joseph Smith get away with proclaiming a revelation through dreams in which he read a secret holy book with the aid of magic glasses before translating it from an unknown ancient language into a then already archaic King James's English? Why can't I have dreams like that? Is it new glasses that I need?

SAVING THE TURTLES

There is a, perhaps apocryphal, story about Bertrand Russell taking questions from the audience after a lecture on the nature of the solar system. An elderly person stood to tell him he had it all wrong: "The Earth is flat and rests on the back of an enormous turtle," she is alleged to have said.

Russell replied, "And what, pray tell, is the turtle standing on?"

"Turtles," she said. "It's turtles all the way down."

However inadvertently, I think of that exchange as a good visual metaphor of classical philosophy. Every layer of inductive reasoning requires yet another, without an end in sight. That seems to be the problem if one insists on knowing a safe answer before the question has been fully framed.

I've written before of my utter frustration with the classical philosophers whose voluminous writings over the past 2,500 years strike me as little more than obscure semantic swordplay. I thought, when I wrote those thoughts, that I alone was bogged down in this problem. That was until I came upon Karl Sigmund's book, *Exact Thinking in Demented Times*. His account of intellectual life in Vienna a century ago reveals that scientists were struggling with the exact same problem then as well. An attempt to restore dialogue between scientists and philosophers resulted in the appointment of Ernst Mach and Ludwig Boltzmann, both distinguished scientists, to chairs in the philosophy department. Their views, like mine, were that only evidence (i.e., deductive reasoning) mattered, and that classical philosophical inductive reasoning was a waste of time.

Thinkers of the day, picking up after Mach, were inclined to reason that inasmuch as every experiment yielded results with margins of error, mathematical exactitude offered a higher level of truth than experimental measurements; therefore, they looked to mathematics to provide the logical foundation for a new value system, which would dispense with so-called turtles. For a while, there was optimism that this could be done. Then, to global dismay, a young mathematician named Kurt Gödel found a proof that mathematics cannot be used to prove the ultimate truth of mathematics.

Straightaway, the mess of words rushed back in to fill the void. There quickly came to be a small academic clique, known as the Vienna Circle, whose members produced many philosophical tomes and lecture courses to get the project back on course. There were extensive debates about the possibility of changing the language of philosophy to require propositions to be stated as hypotheses. Their task was none the easier because, as they were trying to clear their heads, the economy and society of Austria collapsed. Soon thereafter, the Bolsheviks and Fascists were rioting in the streets and some academics fell out with one another. It became clear that science could not be separated from the human condition. To the outsider, it must have seemed that confusion, and not clarity, had gained the upper hand. There was chaos after the Anschluss, as members of the intellectual community either fled for safety, were arrested, committed suicide, or changed sides. The group was atomized, and the clarity of intent lost.

Nonetheless, connections that changed our world were made. Gödel's interactions with Alan Turing led to computers, for example. Karl Popper's ruminations about the falsifiability of scientific hypotheses also comes to mind and forms an intellectual basis for modern experimental science.

Reading this moderately difficult book has at least settled my mind that I didn't miss much when I failed to find substance in the classical philosophers. It is somewhat of a relief to know that the endless layers of turtles can be truncated on the foundation of a verifiable and objective reality. These thoughts take me back to Goethe's Professor Faust, who is struggling with a Biblical translation until he lights upon the line, "*Am Anfang war die Tat.*" (In the beginning was the Deed.) It comes as a relief to know that giants from the past have already cleared a bit of a trail for us to follow and perhaps that is the point of philosophy.

SCIENCE LESSONS
FROM HISTORY

Imagine you were a talking head on tenth century CNN broadcasting out of Baghdad, and their contemporaneous version of Wolf Blitzer asked you to speculate on what may be the future guiding social and religious ethos of the world. I know that it's a stretch to bridge the anachronism but indulge me. You likely would have said, "It has to be Islam. The others don't have a chance. Far-Eastern societies are too far away to be an enduring influence on the Middle East, let alone the West. The Christians are a joke, preoccupied as they are with religious orthodoxy and bias confirmation to the point that eradicating freethinkers pre-empts any energy they might have had for novelty or discovery. Islam has developed technology like no other; algebra has accepted the mathematical concept of zero from Indian sciences, our engineers can now use equations to design elaborate and durable structures and Islamic culture embraces all who seek to advance knowledge. Our studies of the cycles of the stars have given calendars to farmers so they know when to plant for the best crops, and astral charts guide merchants and armies in their travels. Our studies of chemistry have given us soap and cosmetics. Our medical knowledge is second to none." And so on…

All of it true at that time. So, where did things go wrong? In the eleventh century, there appeared a character named Hamid al-Ghazali, who trained in science and would have become one of their best, except that his life was intercepted by a profound mystical experience that rendered him hostile to science. He declared that numbers were the work of the devil because measurement of physical constants tied God's hands against the possibility of his future freedom of action. For instance, if an egg rolls off the table, you could use the occasion to measure the coefficient of gravitational acceleration of falling objects, but God's plan might be that each falling egg would accelerate at a unique velocity chosen for the occasion. This was a literal interpretation of Islam as submission to God's will. Al-Ghazali assessed piety as the final judge over curiosity. His teaching undermined the science of the day by subjugating evidence to ideology, and the Middle Eastern world has never recovered from this self-inflicted ideological injury, despite that he is regarded as the second man in Islamic history, next to Mohamed himself.

I watched the world's response to the Covid-19 pandemic with mounting anxiety. Every newscast became a parade of some admixture of facts, misinformation, and denigration of honest science when facts were not politically convenient. North America was not alone in these deviations into conspiracy theories. Brazil also sidelined experts for daring to support evidence against random notions. The advice of the Center for Disease Control was dismissed by the American President at the time, who used the tragedy to fuel a re-election campaign. Some religious leaders railed against public health best practices and, instead of social distancing, they invoked faith in God's protection. At least one evangelical pastor who did so died of Covid-19, and his congregation had many more cases of the disease. This is the kind of intellectual disorder that repeatedly reduced society to wreckage in the past. Could it do so again? Will ideology once again trump the facts? If skilled people are systematically removed from institutional leaderships, the organizations will stagnate and deteriorate, and it will be the work of a generation to repair the damage. The deterioration of a great nation toward mediocrity is never a pretty thing to watch.

As you are reading this, your telephone rings and it is the modern version of Wolf Blitzer asking to re-interview you and to probe how your prediction of a mere millennium ago could have been so far off the mark. You might care to use the Bishop of Ockham of the fourteenth century for cover. It is said that one day, as the bishop was settled into a chair and waiting for his servant to shave him, he was abruptly seized by an insight that caused him to deflect the approaching razor from his neck, leap out of his chair, and rush to his cell, where he dipped his quill into the ink and wrote, "Do not multiply explanations beyond necessity." In other words, "Don't invoke the will of God if the coefficient of gravity is sufficient to explain the incident of the egg. Keep explanations as simple as required by the observations." This insight disappeared into the Church's archives but it resurfaced centuries later in the writings of Francis Bacon and others as an essential feature of the platform on which western science then developed. Today this insight is known as the Law of Ockham's Razor, and it remains a valid defense against conspiracy theorists and other promotors of alternate ideologies.

So, when you get to explain this on CNN, tell Wolf that the scientific basis of modern society, and much else, will be preserved in some form, so long as individuals insist that facts are allowed to carry the day. The process will not be pretty; careers, perhaps even lives, will be lost in the short term. Intellectual integrity in all matters is an individual responsibility that can never be abrogated.

SIMULATIONS AND REALITY

A few days ago, I listened to a panel of scientists and philosophers debating the proposition that our entire universe may be a computer simulation operated for the inscrutable purposes of a hypothetical higher universe. It was an interesting discussion insofar as it led to the question of whether this could be in any way a testable hypothesis. In order to detect faint signs of an external imprint on any part of our current reality, it seemed we need to view our environment at the highest possible level of resolution, as it were, to see through the level of pixelation that is inherent in the 'upstairs' programming.

The debate held my attention for a while until I tried to visualize the mapping that was under discussion; essentially, on a scale factor greater than 1:1. Umberto Eco, one of my favorite authors, once wrote a piece on the asymptotic approach to uselessness of maps as they approach the scale of the reality they summarize. Maps are useful, in part, because they simplify reality and present only the currently cogent features to the reader. A map as big as Toronto—one that depicts the details of every blob of chewing gum on the sidewalks—will not be useful in the way that a sketch on a scrap of paper in my pocket will be in helping me to get from Queen's Park to the Gardiner Expressway. And it is unlikely to reveal the hand of God upstairs.

But . . . Wait a minute. We are now held in thrall by Google and GPS technology, which provides maps on a sliding scale according to the need of the moment. Gone are the foldout maps of our youth. We now have handheld displays that show the outlines of a thousand-kilometre journey on a screen and that scroll to higher resolutions as we near the destination. A decade back, I travelled with a party to New Delhi; our agent had reserved a taxi for us at the airport, and there it was. But the driver had never heard of our hotel and could not guess its location. We drove away from the airport and stopped frequently to ask directions from passers-by. It took about an hour. Gone are those days, and now every traveller takes control of such situations with their own smart phone.

Given that travel has recently become hazardous in certain parts of the world, perhaps would-be travellers can resort to simulated vacations; they can plot the way to a destination, simulate a flight, and picture the grand hotel and its panoramic vista from the penthouse suite at the end of the journey. Perhaps some undesirable elements of verisimilitude can conveniently be arranged to go missing; we can do without the bone-weary fatigue, the jetlag in a strange airport, the sweatiness of massed crowds, the discovery of bedbugs in one's room, the ever-present risk of food poisoning and of being taken hostage. Still, if you use maps that are scaled 1:1, you wouldn't be able to tell the difference. And if someone can work out a method of scaling maps above that, we may even discover the identities of those beings who may be manipulating us from their upstairs universe.

SMALL MODULAR REACTORS

At the end of WWII, we found ourselves in possession of an energy source that was more powerful and more abundant than any familiar sources of combustion power. Basic calculations told us that there is about 2.4 million times more energy obtained by the fission of U^{235} than from the oxidation of equivalent masses of fossil fuels. Even larger energy yields could be achieved by fusing hydrogen into helium, but that was a challenge that could not then be addressed. There has been some amazing progress toward nuclear fusion energy in recent years, but this presentation is only about fission, the stuff we already have some familiarity with.

The first-generation reactors were primarily research machines. They were used to explore nuclear science and the potential of nuclear power, and to begin testing reactor designs, component, and fuels for their ability to withstand the harsh radiation environments expected to occur in power reactors. They were also used to work out durable approaches to workplace and community safety.

Second-generation reactors were test machines that assessed such features as constructability, optimal technologies, quality of components, and potential consequences of specific forms of failure.

Power reactors in-use now may be thought of as third in the series, and they are broadly accepted as safe and economical. In my own awareness, talk about essential features of fourth-generation reactors began soon after the Chernobyl accident. Since that time, passive safety has been seen as fundamental to next designs. Many other desirable features then also came up for consideration, including those that would increase fuel efficiency and reduce the amount of radioactive waste for long-term storage and disposal.

The reactors now in-use are typically very large and positioned to serve large, concentrated populations. The requirements for environmental assessments, which include case-by-case redesigns to incorporate the most recent innovations and concurrent public education, are slow and cumbersome; financial barriers are huge, and there has been little room for private investment. Size has also contributed to a public misperception that nuclear electricity is too big to be comprehended; therefore, it must be dangerous given the associated heavy duty governmental involvement in regulation. In Canada, where many large areas have small populations, large power plants and extended grids could be more problematic than useful. Small communities will be better served by other means of electrical generation including smaller reactor options.

We have, of course, been building small reactors all along; those that power the US fleet of nuclear submarines and aircraft carriers are "small" relative to land-based power generators. However, "small" isn't the whole issue.

Up to the present time, most installed reactors have used solid fuels configured into rods and sealed in metallic cladding. Fuel rods do not become *waste* in the sense that the uranium has been used up. Rather, the fuels and their cladding swell from neutron bombardment, and the resulting atomic displacements generate pressures that could cause the rods to rupture in prolonged use. Also, the efficiency of the fuel is progressively reduced by the accumulation of fission products.

Heat transfer from the nuclear island is currently done via high pressure steam. Some of the new designs will moderate and cool the fuel with molten salts, and in some of those designs, the fuel will be melted into the salt rather than remain in solid form.

These approaches will enable reactors to be operated at much higher temperature (about three hundred degrees) than a water-cooled system and will obviate the cladding issue, allowing more complete burnup of fuel. Such features will greatly reduce the amount of radioactive waste. There may be about a hundred different models of small reactor products with many variations on offer in the current market. This is certainly a strong testimonial to the creative genius of engineers.

In New Brunswick, two such reactors, dubbed "Waste Burners" are now under construction. Each will each produce 300 megawatts of electricity, and they will be fueled only with already-used fuel from the next-door Canada Deuterium Uranium (CANDU) reactor. They could also be switched to using fresh uranium or thorium, given a compelling reason to do that. Most of Canada's provinces and the Federal Government are contributing to the support of research on these and other models of small modular reactors (SMR), and several (including Ontario and Saskatchewan) have committed to one of these models in their future grids.

One feature of SMRs is that they don't have to be fabricated in the field, as was the case with the currently operating stations. Factory-based construction is envisaged with standard quality assurance issues managed in a controlled environment. Transportation to the site will be on flatbed trucks or by helicopter. This will enable easier provision of electrical and heating services to Arctic and other remote communities.

It is fairly obvious that the ability to place reactors that come preloaded with a decade or more of fuel into remote communities will alter the design and function of existing and future electrical grids. Consider the case of Saskatchewan, a large province with a sparse population where existing infrastructure consists of a coal plant and a distributed network of natural gas-powered facilities, many of which are due for replacement. A large nuclear plant à la Darlington would produce enough electricity to support the province, but the grid to a sparse population would be about as costly to build and maintain as the generating plant itself. An attenuated grid with strategically sited multiple SMRs may meet their needs for the present and provide them with flexibility to grow as required in future.

We have reached a level of sophisticated technology in which there exists a possibility of choices that can flexibly differ from place to place and over time. That may be the essence of maturity in this industry.

SO, YOU WANT TO BECOME A DOCTOR; HOW GREAT IS THAT?

Dear K and M,

Thank you for dropping in to talk to me a few nights ago. You were asking questions about the life of a medical doctor, and I didn't answer them all as well as you deserved. I'm amazed that you are thinking about career possibilities at your young age. Here is my second attempt to give you things to think about.

You asked what it feels like to know that one has saved a life. Let me tell you about one such experience I didn't recall until after you had left with your Gran. It's from my first year after medical school.

I was an intern on duty alone for the night in the Emergency Department at the University Hospital in Saskatoon in 1963. After midnight, we received a phone call concerning an incoming emergency. A farmer had fallen off his tractor while it was in motion, and the wagon that he was towing had passed over his chest, fracturing most of his ribs. When he arrived, it was obvious that his attempts to breathe were exhausting him because his fragmented rib cage collapsed whenever he tried to draw breath. At that time, ventilators had not yet come into clinical use. What can one doctor working alone through the night do to help? I had heard discussions of flail chests but never seen one before. The thing I thought to do was to force some surgical towel clips through his skin into the collapsing ribs, and then attach a heavy cord to those clips. I ran the cord over the top of an IV pole, and added weights as needed to pull the chest wall up. This eased his breathing difficulties, and he was stable when he was transferred to the surgical ward.

The next day, I heard that the Professor had been around to see the patient, had marveled at my contraption, and had told the people with him that I had saved the man's life. Of course, I was gratified by his comments, and it made me more confident about myself as I went about my other duties that day.

Our discussion the other evening went on to other facets of life-saving interventions, and I pointed out that preventing harm may not seem as glamorous as facing into a disaster when it happens, but prevention is also essential and important work. For instance, the person who spends time discussing the dangers of cigarette smoking with teenagers may lead a young person not to take up smoking today, and that may save a life years from now. I want to extend that thought a bit more with an example of something I learned last week, while it was my turn to be a patient for a few hours.

I was reading a book about the links between social conditions and community health, including epidemics, while I was in a hospital waiting room. In the nineteenth century, European cities were generally filthy, and disease levels were high. There were no sanitation systems. Ditches outside the houses were full of dirty water that didn't drain, and everywhere in these cities there was a foul stench. Edwin Chadwick was an engineer who thought it would be good to build houses with a water supply coming directly into each house so that people could have clean water for drinking, cooking, washing, and then flush away waste into a sewage system. Then, another British inventor (whose name was Thomas Crapper) produced the first flush toilet. After that, people began to appreciate that the situation would be improved even further if they cleaned up their homes and their yards. The evil smells gradually went away, and people generally felt better. The death rate from infectious diseases in those cities also fell very sharply. Neither Chadwick nor Crapper were doctors, but their inventions saved millions of lives in Britain alone and continued to do so after their deaths and to this day!

My point is that you don't have to become doctors if saving lives is your goal; it is only necessary that you find a job that you love doing and use your education to learn to do the job well. I wish you success in all your adventures.

SOS

It seems that Kenya's recent Minister of the Environment took her job seriously. That country has now banned plastic bags with penalties including jail time and fines of up to $49,000 (CAD).

The news item reminded me of a drive across the Sudanese Sahara Desert in 2003 where, after several hours of driving along a barely marked trail on the sand, we came upon a lone thorn tree that looked as desolate as if it had been abandoned by the Creator Himself. With nothing else about to suggest even the possibility of life, and to complete the desolation, the tree was completely festooned by plastic bags that the hot winds from Hell had impaled on its thorns.

When I first visited the Old Gold Souk in Sana'a, the entire marketplace was surfaced by discarded and broken plastic water bottle remnants that weren't being cleaned up. Under the bombardment of the heavy foot traffic, the bottles had broken up into small, mostly sub-centimetre plastic flakes. Nearby to the market was a wadi (a dry riverbed) into which the runoff flooded during the rainy season to form a temporary river, and it carried away much of this detritus. To the credit of the Yemeni, this mess was cleaned up before the celebration of the tenth anniversary of the country's unification and, thereafter, there were uniformed cleaners moving through the crowds, picking up trash. One was left to wonder whether their gatherings were not simply dumped somewhere at the margins of the city, only to be washed into the wadi along another path during the next rainy season, but it may have been a move toward a right direction. I haven't been back for more than a decade, but I don't expect that the years of civil war since have improved public environmental awareness there.

My first visit to Indonesia was for the purpose of attending a medical conference that took place at an elegant hotel on the island of Bali; not even a blade of grass out of place. Eventually, I felt a need to briefly get away from the stresses of the formalities and took a walk along a path that led to a nearby village. The walk was abruptly terminated when I came upon an open, blocked street sewer that ran by a line of homes. Lots of dead chickens were afloat with other noxious debris, as well as water bottles and a stench I recall as the worst I've ever experienced. In the intervening thirty years, the waste disposal problem in Indonesia has only become worse. As we approached the country by night on a luxury cruise ship about six years ago, the vessel ran into a floating raft of garbage that blocked the cooling intakes and stopped the engines for a time until the crew could clamber overboard in the darkness and clear away the debris. In a subsequent visit to a busy harbour, we walked through a suburb of homes that were no more than platforms on poles suspended over water along the shoreline, and where all forms of garbage, including water bottles, once tossed, were floating out to sea. Most recently, there was a news item noting that the Indonesian army was being used in an emergency action to clear blocked streams and rivers across the country where water bottles and other debris were damming the drainage.

I am sad to say that Canada is also a contributor to the problem. In 2013, the harbour authorities in Manila stopped the dumping of 2,300 tons of unsorted household waste from Vancouver that had been falsely labelled as recyclable plastics. Until 2017, the ship was kept docked there with its full load of rotting food waste and soiled diapers. The matter was resolved with the ship's return to Canada, still fully loaded and the cleanup rightfully fell to us.

In another story on CBC Marketplace, we learned that ninety-nine percent of Canadian donated clothing ends up in the trash, and we were shown full containerloads of the same being incinerated in Kenyan landfills. It turns out that the average Canadian wears an item only six times before discarding it. Are we not doing laundry anymore?

Where will this litany take us except into the depths of despair? My own former city, Ingersoll, long debated the merits of a proposal to redevelop a local exhausted limestone quarry into a landfill. The project had many intriguing aspects, both technical and concerning regional planning. There was the need to protect the groundwaters from contamination and the city from vermin, odours, traffic issues, and the like. On the planning side there were also questions such as whether a landfill was the best possible use of the quarry and whether it would attract unwanted wastes from adjacent municipalities. These are all good questions, but vocal emotional motivations deteriorated the debate into a dogfight in which evidence counted for no more than the hysterical opinions. Some politicians used the situation to attract votes.

Our waste management problems worldwide seem to me to be but one more example of the old problem of the commons. The resources and responsibilities that belong to everyone, in fact, are cared for by no one. Many of our public policy decision makers are intellectually imprisoned in ideologies that define them as either capitalists or socialists, neither of which can solve all our problems. I think of them as home repairers who might brag that they use only duct tape, never a nail, to complete their tasks. Let us not define our approaches to solutions before we have understood the problems. And, be assured that this problem of waste management urgently needs a comprehensive long-term solution.

STIRRING MURKY WATERS

We don't know how life on Earth began, which makes the topic an object of endlessly creative speculation. During the pandemic, I occupied myself in part with Eric Smith and Harold Morowitz's recent book, *The Origin and Nature of Life on Earth*. In many ways, this is a book that wouldn't hold my attention if I had something better to do, but the authors have laid out a huge framework of evidence for their speculations, such that they are not to be dismissed out of hand. Their basic thesis is that before there could be life, there had to be a pre-biotic chemical evolution. Various bits of future biochemistry "had to experiment" with the best ways for the planet's high energy electrons to cascade toward a stable low energy state while following the fundamental laws of thermodynamics that Ludwig Boltzmann deciphered in the late nineteenth century.

An urgent early question concerns how a chemical reaction occurring in water could be maintained in a high enough concentration without loss by dilution to kickstart the next step in the reaction chain. It seems to me that this tenuous step could have been protected in one of two ways. The first would have been to allow the earliest proto-life reactions to occur on the surfaces of crystals. This solution could have contributed to the resolution of several other early life enigmas as well, including the "handedness" of life, the fact that all of life on Earth uses only those sugars that rotate polarized light to the right and amino acids to the left.

There are mineral crystals that share a similar handedness and may have impressed their trait onto the products of the first proto-enzymic products. Crystal surfaces may also be an explanation of the observation that many of the critical enzymes that drive modern metabolism have mineral atoms in their structure: calcium, iron, magnesium, manganese, zinc, etc. Possibly the persistence of these elements in the active centres of enzymes today is a nod to those early times when proto enzymes were hanging onto a rock face, using what they found there to struggle forward and they carried some features, including the metal ions, into modern molecular structures.

Another way to contain the nascent biological reaction in sufficient concentration is to contain it in a membrane. Someone has provocatively described cell membranes as "life's space suit on Earth." Who supplied the space suits? It turns out that the raw materials of which the planet is made contains all the materials needed to make the molecules that are essential for life. There is spectroscopic evidence of fatty acids in the supernovae from whose dust we were made; there is direct evidence of fatty acids in meteorites striking the Earth. In water, these same molecules spontaneously assemble to form microscopic vesicles that bear a close resemblance to modern cell membranes and soap bubbles. Studies of complex structures have become very interesting as many complex structures are being found that self assemble from simpler precursors.

Another feature of fatty acids is that they are electrically asymmetrical. In their spontaneous arrangement, they form membranes with the fat-loving ends on one side and the water-loving on the other. This configuration takes up energy from electrons and, indeed, every cell membrane carries a charge of about -70mV. Thus, a cell membrane is an electrical ground, and its metabolic machinery is the dynamo that maintains the charge.

Finally, for brevity's sake, crystal rock faces may have preceded the physical appearance of membranes. Much metabolism occurs on the surfaces of membranes and not in free solution in the cell. Think of the mitochondrion where the enzymes that break down sugar to release energy into ATP are firmly fastened to the membrane, and the only thing that moves is the sugar fragment, like a car along an assembly line. Other metabolic pathways are found on the cell membrane, the nuclear membrane, the microsomes, the Golgi apparatus, and the endoplasmic reticulum. The membranes themselves seem to resemble the face of a crystal. This is also characteristic of the nucleic acids (ie, RNA and DNA), except that they are working with a one-dimensioned membrane (ie, a string). Very little of the machinery of life operates in free solution.

Two centuries ago, people who thought about the origin of life divided into two groups. There were the Mechanists, who looked for the entire answer to pop out of a test tube, and the Vitalists, who insisted that the transition from non-life required an intervention from a higher order. In modern times it is evident that they both missed something of the beauty of the chemical evolution that had to precede the commencement of biology.

STUFF THAT GOES
BUMP IN THE DARK

Physicists know they are in perilous territory when an equation yields an infinity. That is the case for the density of a black hole; a mass divided by zero volume equals infinity. What does that even mean?

I would have liked to have met Karl Schwarzschild, an early 20th century German physicist who found himself serving in an artillery battalion during the First World War. Somehow, he mustered the intellectual stamina between incoming and outgoing barrages to write a landmark paper about the theory of black holes. The paper has stood until today as a sentinel achievement, in which he described the effective radius, known today as the Schwarzschild radius, at which matter, and even light photons cannot avoid being drawn by gravity into the hole. This distance has come to be known as the Schwarzschild radius; ironically, his name means "black shield," which in a way is what his radius is.

Black holes come mostly in two size ranges. The stellar variety are the small ones that result from the collapse of stars of about eight–ten solar masses. Then there are the very massive black holes that are found at the centres of spiral galaxies and measure millions to billions of solar masses. It is a mystery why there seems to be so little in the middle between these extremes, but it does suggest that the two types have very different histories of formation.

There is not much that we can detect from Earth concerning black holes beyond their gravitational fields and their spin. They exert a microlensing effect, which is the distortion of space by gravity that bends the path of a light beam as it passes near a black hole. Einstein himself did not hold out hope that gravitational waves would ever be detected, despite his having shown their theoretical existence. That pessimism didn't stop later generations from trying. After a half century of trying, success came in 2016, when the Laser Interferometric Gravitation Observatory (LIGO) was completed. The project was the culmination of four decades of effort involving several insightful scientists who, with about a thousand collaborators, developed the essential technology.

The LIGO laboratory consists of two identical installations on the east and west American coasts so the expected small signals would not be swamped by noise. Signal detection depends on recognition of mirror displacements of less than the diameter of a single proton ($<10^{-20}$ centimetres) over four kilometres.

As the installation neared completion, the staff and students were occupied by final calibration measurements and there are several stories told about the first detection. Ray Weiss, a senior scientist, gave instructions that there would make one final calibration run without actual data collection. The students nodded and winked because they knew they could satisfy his requirements while also collecting data and someone flipped the switch without telling the boss. The signal from the collision of two black holes came within the first hour. The findings were kept under wraps for several months while every aspect of the event was checked out and characterized. They had to be certain that they were dealing with a true signal. It was finally determined that they had observed the collision of a pair of black holes with respective solar masses of thirty-six and twenty-nine, at a distance of 1.2 billion light-years. The impact resulted in the conversion of three solar masses of these black holes into energy, and the emerging new black hole from the crash site had a mass of only sixty-two solar masses. This observation netted a Nobel Prize for the three most senior scientists.

The lab went on to detect eight to ten more collisions over the first year, and then the interferometer was taken offline for a detector upgrade. In the second run, they had much improved sensitivity, and more than sixty additional events were recorded, many involving smaller objects, including neutron stars. It has been upgraded again, and it is reported that several collisions are being recorded each day. A similar detector called Virgo has been established in Italy, and others are on the design board for India and Japan.

The Europeans are developing a Laser Interferometer Space Antenna (LISA) that will be deployed in space, with detectors placed more than a million kilometres apart. There is also an international collaboration that has reported success from using disturbances in the timing of signals from rapidly spinning neutron stars as a detector of gravitational waves. The most recent report I saw featured nearly 60 neutron stars in a coordinated network and confirming the feasibility of this method. The theoretical expectation is that there are up to six million stellar collision events per year waiting to be detected. I wonder whether there is any hope of ever detecting a gravitational wave from the Big Bang itself.

Interferometers do not focus well enough to give accurate localizations of collisions in sky coordinates by themselves; the best we have now is overlapping banana-shaped probabilistic patches. In the future, and facilitated by the opening of additional installations, it will be easier to back triangulate all the observations to smaller sky areas and bring telescopes into coordinated play. The use of both gravitational forces and electromagnetic radiations promises to be highly informative on scales not seen before.

And you may have thought that the universe had no more secrets to reveal! Masses of infinite density and zero volume are hiding behind black holes and smashing into each other with some regularity. How weird is that? The universe feels at times like a cosmic paintball battle/funhouse. Actually, it's better than that. Einstein would have been impressed.

SURFING OVER HISTORY

In ancient Greece, Democritus advanced an argument for the existence of atoms as the smallest irreducible quantity of a substance. However, the honour of advancing atoms as the basis for a modern theory of matter was reserved for John Dalton in 1803. Atomic theory became a useful hypothesis then because it provided the way toward measuring the relative weights of the elements and their propensity for proportional combination with other elements to make compounds. Everybody who was anybody was into science in those days, and Amedeo Avogadro, a lawyer who became Italy's first professor of physics, propounded that a standard volume of any elemental gas should contain the same number of atoms at constant temperature and pressure. Avogadro's number has since been burned into every chemistry student's brain: 6.022×10^{23}. Still, people were not universally convinced of the actual existence of something so small as to ensure that it would never be seen.

As late as the beginning of the twentieth century, there were still debates featuring giants of science, as was the case between Ernst Mach and Ludwig Boltzmann. Soon thereafter, Einstein explained atomic recoils as the basis for the billiard game represented by Brownian movement, such as you can see if pollen grains suspended in a drop of water are viewed under a microscope. Still, I recall when, as a child, I was reading my way through the *Encyclopedia Britannica*, the entry on atoms cautioned: "Don't take all this too seriously. It may be nothing more than a useful fiction." Since then, we have seen atoms by electron microscopy and counted all their constituents. Atoms exist. True, we do not yet have a method by which to visualize quarks or to confirm that they may be broken down into strings. Nevertheless, atoms exist, and Democritus would have been content.

In Charles Darwin's time, and for some decades after, there was no physical basis to support speculations concerning the peculiar whimsies of biological inheritance. That began to change with Gregor Mendel, who noted that crossbreeding peas of different colours didn't yield progeny of smeared colours, but only of the original hues in definite arithmetic ratios. Thus, was born the concept that inheritance came in discrete bundles of genetic information—i.e., genes. The reality of how genes might physically be constructed remained totally obscure for another half century. Little by little, degrees of clarity emerged; George Beadle and Edward Tatum performed many experiments with bread mold (*Neurospora*) and showed that mutants sometimes had nutritional requirements that had not been present in the parent strain. Eventually they demonstrated that each nutritionally deficient strain lacked an enzyme that was required for the synthesis of an essential amino acid—hence their enunciation of the "one gene, one enzyme" hypothesis in 1941. At that time, the gene remained little more than a concept, but the route toward resolving that mystery became clear twelve years later through the work of Thomas Watson and Francis Crick on DNA structure. Today we have a deep and still expanding understanding of genes, of their structure, their relation to our identities, and what they do or do not do for us, as well as the nature of environmental controls that act to adapt us to our world. Nonetheless, much remains to be learned, and there may yet be another century of work before we will really understand DNA.

I am deeply curious about what makes us fascinated with certain problems, beginning with speculations, moving toward ever-deepening pools of hypotheses, and, through observation and experiment, onto the harsh rocky shores of evidence. These three anecdotes all point in the same direction, though they involved thinkers living in different societies with unique views on the nature of the world and puzzling different problems. At the outset of each of the foregoing anecdotes, there seemed to be nothing more at stake than interesting speculation sufficient to spark conversation in a Greek bath, a Viennese social circle, or an English pub—something interesting but forgettable in the aftermath of real life. But speculations sometimes have consequences and grow up to become basic to our evolving societal underpinnings; that is why education is cost-effective. What began as speculation may snowball toward testable hypothesis, and eventually, to reveal evidence.

There is often a lapse of time in the progression from intriguing observation until the appearance of a testable hypothesis—time that is consumed by learning how to ask the question. There is danger lurking in that interval where diverging speculations can harden into convictions that are no longer subject to amendment by new observations. That seems to have happened in the creationist versus evolutionist debates.

We may be in that situation today as regarding the nature of dark matter and, even more so, of dark energy. There is a prevailing hypothesis that dark matter should be particulate and describable within the standard model of matter, but no candidate particle has yet been identified. Are we asking an answerable question? Dark energy is even more problematic, as we seem not to know even how to ask. There is not so much controversy here as simple befuddlement. Once more, the law of energy/matter conservation seems to be in jeopardy, and we may have to cope with that as a consequence of new concepts that challenge thinking about this subject.

I'm not doubting that our current boundaries of ignorance will be overcome; it will take time, and some of us will have sleepless nights on the way there. I take heart that we are supported by millennia of careful thinkers who built durable foundations. Uncertainty of outcome is the burden we accepted when we chose to exercise our curiosity and most of the time it is an exciting journey we are on.

TELLING THE TRUTH

While I believe passionately in the existence of truth, I also acknowledge that human ability to recognize it will always be limited by the sensitivity and specificity of measuring tools that set the error bars on all our observations. There will always be a possibility of error, and there will always be a price to pay for that. Evidence-based thinking has been the arrow pointing in the direction of truth for the past four centuries and, despite error bars, has enabled the compounding advances in science that now subtend our improving quality of life. But how should we respond to our errors when they become evident?

In the past fifty years, more than one hundred million scientific articles have been published. Most authors do their best to verify their data and control their speculations. Scientific journal editors provide an independent check on opinions and speculations that go beyond the data, largely by submitting all manuscripts to peer review. Manuscripts of dubious quality are routinely rejected but the process is not efficient at detecting error, fraud, or plagiarism. Until recently, about forty percent of retractions were at the initiative of the authors themselves, and these humbling admissions are a credit to their integrity. These are the easy bits around the fringes of the problem. There are many settings in which the organizational and performance pressure on scientists prompts conflicts, which come about by ethical and human tensions.

Other pressures also arise. What is the proper course of action when an established investigator's experiments don't replicate in other laboratories?

A decade ago, there was a lot of excitement as the new concept of primordial stem cells existing in adult heart tissue was announced by Professor Piero Anverso. Many universities spent a lot of money to set up competing labs, but then the work was unreproducible. Anverso was then found to have invented his critical "findings." Harvard University was forced to close several labs and terminated the careers of some scientists. The misdirection that preceded the identification of fraud wasted a lot of resources and the valuable time of many career scientists.

On another tack, even if the data are correct, there can be problems around the ethical dimensions of the work. For instance, the Retraction Watch site noted that a stream of publications concerning organ transplants from China seemed to have utilized organs from executed criminals; they expressed uncertainty whether the demand for organs was driving the sentencing to executions. Efforts were ongoing to retract the offending publications when I read the report.

A recent development concerns the appearance of paper mills, fraudulent publications often based in Russia or China that may use artificial intelligence to develop plausible articles in which they may offer to sell authorships that would help to fatten a thin academic CV.

Lest we be tempted to rush forward with the mob and brand all science as "fake news," let us first remember that science is a product of fallible human minds and, as such, it is subject to correction when evidence requires that to be done. There is a t-shirt mantra that advises "Ask, Test, Repeat." I would further note that with each such cycle, one is likely to alter the question a bit until the answerable form of the query emerges. The majority of scientific publications continue to report on science that was developed and reported honestly. That is the only way we can screw ourselves forward through the matrix of ignorance in which we are embedded.

THE AFRICAN SKY

The central African night sky is the most heart-stopping view of the sky I've ever seen in my life. At that moment, my grandson, Luke, and I were somewhere in the middle of Botswana, walking across the dark clearing of our encampment toward our dinner. As we cleared the silhouette of the last tree in our path, a large sector of sky came into view and abruptly, and we had a full view of the Milky Way. There were the Magellanic Clouds, two of the Milky Way's fifty satellite galaxies. In its proper place were the Southern Cross and Orion presiding over the top of the galactic arch. The stars were shining brighter than dagger points in the dry air, and without light pollution. I wasn't ready for the spectacular view; it was awesome. Later that night and in our tent, we were awakened by an elephant who spent some time trying to uproot the tree to which it was secured. It was a time not to be forgotten.

I can understand those ancient Greek shepherd boys who were mandated to sleep nights at the gate of the village sheepfold to keep away the wolves, and who bravely told themselves bedtime stories of the stellar constellations overhead that warped under the influence of their lonesome, hormonally confused brains into beautiful women, and even goddesses, whose stage was the sky. "Look up there. That's Cassiopeia." For our own reasons and wild imaginings, the night sky remains magical. I try to understand the mind of Giordano Brno in Rome, who marveled at the stars and was overheard to speculate that each one of them might, just possibly, be a replica of our Sun, complete with planets and lively populations of its own, a speculation that cost him his life. He was burned at the stake in 1600. Today we respect him for his courage in thinking outside the theological constraints of his time's box. We know now that he was at least partially right; the stars are suns with planets, and the very powerful telescopes we are now building may give us further glimpses into the question of extraterrestrial life within a decade or two. The anticipation of what new information may soon come has me breathless with anticipation. This is the deep magic of a sort in the night sky and it's not a trick.

Galileo's invention of the telescope signaled the end of the medieval Dark Ages. He came close to sharing Brno's fate, but his sentence was commuted to lifetime house arrest because he and the Pope had been childhood friends. His telescope showed that the Earth is not at the centre of all movement in the universe, thereby supporting the heretical views of Copernicus. He also observed that there are moons in orbit around Jupiter and that our moon, far from being a perfect body in heavenly space, had been the victim of many cratering impacts. Simply put, Galileo proved that Aristotelian cosmology, heretofore credited with the authority of infallibility equal to that of the Scriptures, was deeply flawed.

It's all there in the African sky, and I invite you to put that view on your life's bucket list to see. Short of that, there are great images from the best, currently operating telescopes on the Internet; there are many new ones from the James Webb Space Telescope, including images of some of the first stars ever formed. If you have the time, wait a little longer for the Vera C. Rubin Observatory to come online in 2024, which promises to reveal about a thousand supernovae every night. (In Kepler's time, they could see about one per century.) The images are sure to be awesome, and they will challenge us to ponder again how it all, indeed, how we all came to be.

THE BARON MUNCHAUSEN

I was reminded this morning of a character in my grade school reader, the Baron Munchausen, who had a gift for rotund exaggeration. He seemed to like hunting stories such as the one in which he supposedly positioned his hunting knife to accurately split his last bullet into two fragments and, thereby, bagged two prize bucks with one carefully triangulated shot. I suspect that the Munchausen archetype was drawn from a then-common malady, now remembered as neurosyphilis, which stereotypically resulted in a form of dementia associated with posturing and pretentious grandiosity. In evidence thereof, when I was in medical school about sixty-five years ago, there was an inmate at the Prince Albert Sanitorium who liked to talk to medical students and believed he was Jesus Christ. The advent of penicillin ended the tragedy of neurosyphilis and collapsed it into the medical history books, but the Baron lives on in other ways.

I ran into him again in medical practice, but he was wearing a new disguise; his name became attached to self-inflicted, attention-seeking behaviours that were intended to mislead physicians and other care givers. There is a medical literature about this form of the Munchausen syndrome.

Early on in my practice, there was a person who had learned that a lot of attention could be gained by presenting at the Emergency department with complaints of chest pain and coughing of blood. This complaint had to be fully investigated because the possibility of a deadly blood clot (embolism) had to be ruled out in every case, including by the performance of chest x-rays, blood gases, and lung scans, all of which would prove to be normal. These patients recycled and used all the hospitals in the region to avoid detection for as long as possible. In that era of paper medical records, the communication gap between hospitals made it difficult to run the problem to ground. We were concerned about this in nuclear medicine because the lung scan of that time delivered nearly a hundred times more radiation than it does today, and we wanted to avoid the possibility of doing harm by over-investigation. One day, a student rotating between hospitals identified for me a patient as one who had presented elsewhere a short time earlier with the same story. I made the extra effort to obtain all the records from the region and verified that this person had been repeating their fabrication at the regional hospitals for a while.

Electronic records today make it a trivial matter to suss out such attempts at deception. At the same time, technology also changed the presentations of the Munchausen syndrome. They now turn up on Facebook and in various chatrooms, where they misidentify themselves as cancer patients who have many other chronic or incurable diseases that incite empathy and contributions to GoFundMe pages. We could say that the meme is mutating! Where these once presented mostly in doctors' offices and mainly endangered themselves while seeking out treatment for diseases they didn't have, they currently have audiences to whom they can display themselves, and that will have consequences for others. In short, in their modern guise, they may not any longer be a danger only to themselves. If only the Baron had stuck to telling amusing stories about his hunting prowess.

ORIGINS OF LIFE

Unlike Athena, who was said by the Greeks to have sprung fully grown from the brow of Zeus, I think that life did not first appear on Earth in any of the forms that it takes today. Indeed, there may not have been a clean separation between life and nonlife at the beginning—there was perhaps something more like a gradient of states that gradually shaded toward life. For instance, there might have been an initial self-replicating enzyme that was capable of mutating. For the entertainment of those who imagine that life is too complex to have had a spontaneous origin, here are a few facts (with speculations to follow later).

Let's begin with a line of prebiotic chemistry learning that comes from infrared spectroscopy of supernovae. When a star dies, the energy released in the supernova explosion results in the *de novo* creation of heavier elements of the periodic table beyond hydrogen and helium. First, carbon is created, along with nitrogen, oxygen and more elements up to iron. These cosmic clouds of highly reactive materials produce at least hundreds of organic molecules, many of which are essential for life as we know it, including fatty acids, amino acids, sugars, alcohols, nucleic acid bases, and many others. We can detect these compounds by means of chemical spectroscopy conducted through visible-spectrum and infrared telescopes. Telescopes in space, such as the James Webb Space Telescope, are best suited to this task because infrared frequencies are absorbed in our atmosphere. Over the ensuing eons, this cosmic chemical debris will eventually condense with interstellar dust to form planets around the next generation of stars. This happened on Earth about 4.6 billion years ago.

We know that life-related compounds are synthesized in space because they are found in meteorites that were formed in space before our planet existed and from the same debris that later fell to Earth. In this way, our planet was impregnated from its beginning by the essential chemicals that are part of the "starter kit" for life. These insights, coupled with the now-proven existence of planets around most stars, raises the compelling question as to whether any form of life may also exist on some of those planets. How different will it be from anything that ever existed on Earth? Given the nature of the cosmic seed, perhaps not very different. We are not necessarily unrelated to entities, even on distant exoplanets.

Think about it: every atom in our bodies came from the bellies of stars that exploded more than five billion years ago! We really are made of stardust—i.e., recycled stardust at that. It appears that all planets may have been seeded with the same prerequisites of life and that our Earth is not unique in the universe.

THE CASE OF THE CARDINAL
AND THE COLLEGE

The image I wanted to display may not be online any longer, but a dimorphic cardinal, male on the right and female on the left, will be found if you enter the term "dimorphic cardinal". How can this be? One wonders if the cardinal is similarly confused. I assure you that photoshopping was not involved. This bird had better not show up in Republican states because nonbinary entities are rapidly becoming illegal there.

Seriously, this is a case in which the female parent cardinal laid a double-yoked egg that was supposed to become two birds—one male, one female. Within the limited confines of the shared eggshell, and at a very early stage of the incubation, the two embryos touched and fused into a single germinal entity. Then the Hox genes that control the head-to-tail and left-to-right layout of the body took over, using what material they were given to work with, and they made this confusing creature. The Hox genes are universal and exist in all life on Earth with little, if any, variation. At the proper embryological time, they also controlled our body layouts.

To go on with my story, I will have to introduce some additional elements of mammalian physiology and even theology: from the cardinals to The College of Cardinals, if you wish.

As a young physician, I was approached by a woman who was not at peace with her body. "Why," she asked, "am I a redhead on one side of my body and a brunette on the other? Why do I have eyes of two colours?" I explained that in her mother's womb she had been on a path toward becoming twin girls when the two eggs came into direct contact and stuck, exactly as the cardinal did in its egg. If we had had DNA sequencing for my patient back then, the report would have stated that this DNA sample was a mixture from two sisters.

Years later, I attended an exhibit of fourteenth-century Florentine religious art at the Art Gallery of Ontario (AGO). The story board was about the theological debates concerning the origin of life. I was there to see the wood carvings, but I learned so much more in passing.

Not knowing about mammalian eggs, those 14th century scholars all believed that the mother's contribution to the generation of offspring was no more than to be the vehicle that incubated the father's seed. And all agreed that life began at the ensoulment of the conceptus—ie, immediately on contact and as an integral part of procreation. No room here for confusion because the beginning was the work of God, and all His acts were perfect. The dissenting view that development of the soul might be progressive along with that of the body was pushed aside, because it was the understanding that perfection, once achieved, could not be improved upon by changing any part of it.

This view was reflected in the art that I viewed at the AGO that day and it formed the basis of the Church's arguments that, to this day, forbids abortions. It seems that modern biology has not promoted a rethinking of those dated views, although seven hundred years have passed. Be that as it may, whatever does the doctrine say about the occasion when nascent twins fuse in utero to form one? Did my patient have two souls? Does God have a recycling department for supernumerary souls, or do some of us get away with having two? Once more, the Bishop of Ockham, who helped to lay the basis for the Enlightenment and the scientific revolution that followed three centuries later, had a straightforward approach to these situations: he said that you should tentatively adopt the simplest solution that takes care of all the observed factors as it is the one most likely to be correct. He might have wondered whether the soul was even an essential component of these considerations; perhaps he was getting too close for comfort, as he was subsequently excommunicated.

If you should ever happen to speak aloud along these lines and find yourself under interrogation over the source of this heresy, tell them that a little bird told you.

THE COUNTRY HOSPITAL

During our undergraduate training, we were required to spend time shadowing a country doctor, who was a general practitioner in the then full meaning of the term. The time was approximately 1962, and the concept of restricting a doctor's license to areas of demonstrated competence had not yet fully emerged. I was terrified at the prospect of being expected to perform appendectomies, tonsillectomies, hysterectomies, and Caesarean sections, along with caring for a quota of strokes and heart attacks as they came along based only on what I had learned in medical school so far. I already knew with absolute certainty that rural general practice was not to be my life. Most of my class must have experienced a similar measure of insecurity as many of them also eventually sought refuge in some form of specialty training.

My externship took place in the small town of Elrose, a farming community located about one hundred kilometres from my birthplace in western Saskatchewan. The population of about seven hundred was served by a single doctor who was an ideal physician for his community in that he and his clinic blended well with the farm implement dealerships and other mainly agricultural support businesses on Main Street. The province supported his practice by funding a small hospital in the town. The 'hospital' was a remodeled large, once elegant, home of a previous generation in which the dining room had been reconfigured into a delivery suite and the front room as an operating theatre. There were a couple of former family bedrooms upstairs that now served as an inpatient ward. Nursing and other required staffing was largely provided by local farm wives with nursing school certificates. There was a generous amount of multitasking amongst the staff in that the cook might also be the financial manager, purchasing agent, janitor, etc. On a few occasions, a second doctor would be required, as for surgery or in an emergency, and then it became a matter of finding an available colleague in a neighbouring town who could leave whatever they were doing at that hour to drive thirty to forty kilometres over gravel-surfaced roads and get there on time to still be relevant. Ambulance transport did not exist then outside of the cities. I noted when, shortly after my preceptorship, my preceptor also opted to come back to the University for specialty training. It was a challenging practice.

My rejection of that rural medicine lifestyle as a career option is not intended to belittle the contribution that solo country doctors made to the quality of life on the prairies in those and preceding years. In the early 1920s, Saskatchewan had one of the highest maternal mortality rates in Canada—perhaps in the developing world. As I recall, there were about 1,200 deaths in a population experiencing 25,000 births each year. The introduction of these small community hospitals, along with a rudimentary program of prenatal care, brought the mortality sharply down. As more physicians came along, the once lone doctor, whose solo intervention had ofttimes not been enough to save the day, was no longer left alone without collegial support and having to suffer tragedies that extra hands could have helped to avoid.

In my childhood hometown around the time of my birth, every taxpayer was invoiced in the amount of one dollar per year to provide a base salary for a doctor in the community and, in exchange for this, the doctor did not charge office visit fees. These were the desperate community roots of socialized medicine coming from the bottom up. It was a model of care from which all Canadians eventually came to benefit. Those country doctors, limited and imperfect as they were, were better than they ever knew for the roles they played in sustaining that process to its maturity.

At the conclusion of my externship, I wrote a fairly critical report to the Dean about my externship as a wasted experience but, time has brought perspective, and now in my dotage, I am grateful. I still think that I could never have become the equal of those country doctors, and I am thankful that I found other useful employment in which there was space enough to also be a dreamer.

THE COVID-19 VACCINE

As I write this, the first vaccines for Covid-19 are being administered; the rollout began in Britain on Tuesday, and it will soon also begin in Canada. President Donald Trump has issued an Executive Order to limit export of the vaccine from the United States until all Americans have been immunized. At the same time, there are daily sceptical comments being passed around to the effect that the short time elapsed from the beginning of the pandemic to now is prima facie evidence that shortcuts and compromises have been taken to produce a vaccine so quickly. Surely, they imply that the product is in some way unsafe. It's not hard to rebut the sceptics.

First, not all vaccines are similar, and the pathways that led to production of other vaccines didn't all apply in this case. Indeed, in the Covid-19 case, the competition has been severe, and there have been at least 250 independent research programs that have so far produced about a dozen products with a chance of securing safety approvals soon. It's hard to tell whether that includes the Russian or Chinese products. The coronaviruses have their genetic programming written on RNA rather than DNA, and that makes them unique. The chemical techniques for managing DNA and RNA chemistry have been revolutionized several times in the past fifty years; the reactions and analyses that took years to complete when I was a graduate student are now the work of minutes. Katalin Karikó, now medical vice president of Moderna Corp, lost her academic position at the University of Wisconsin thirty years ago because she was too far ahead of her time. The result is we now have the technology that she developed back then, and that no previous vaccine development team ever enjoyed.

Secondly is the matter of financial support from governments. The politicization of the pandemic has served to spur funding. If you want to talk about the "China virus," let's also talk about the "American." or "German," or "You name it" vaccine. There has certainly been competition. And the fast-moving competition has raised speculations about cheating: who has been cutting corners? Who would dare to falsify data when the consequences would be obvious within weeks with direct tracking to individuals with felony charges in hot pursuit?

And that brings me to the matter of vaccine approvals by governmental health regulators. What are the pressures on the national regulators? Has Health Canada rendered a careful and reasoned opinion based on evidence, or has there been pressure applied by politicians or companies to make a non-scientific decision to deploy? How is the average citizen to know who to trust? I'm reminded of the events following the approval of Thalidomide sixty years ago. At that time, there were few powerful drugs in medicine's armory, and regulations on introduction of new drugs were relatively lax. A German drug manufacturer introduced Thalidomide, as a new sedative and for treatment of nausea associated with pregnancy. In Canada, the regulators approved it without in an in-depth review. It was, ironically, a Canadian doctor working for the US Food and Drug Administration, named Frances Kelsey, who suspected a link to the recent increase in babies with multiple birth defects, principally defects of failure of development of limb buds. She blocked Thalidomide's release in the USA and averted further tragedy.

The enduring public lesson from this episode was that there had to be a clear separation between manufacturers' promotions of new products and regulatory approvals by a public service that was equipped to conduct extensive reviews in the public interest. Gone are the cozy days of "Let's talk this out over lunch" between corporate forces and public servants. Heightened standards of conduct and trails of accountability in and out of the office became the new norm.

In short, I believe we have an effective and accountable regulatory system for the introduction of drugs, in which regulators are proudly professional in the discharge of their duties and are also aware that failure to follow all protocols will have consequences. In my own career, there were times when the ponderous deliberations of Health Canada were frustrating, but from a little distance, I can understand and agree with their insistence on following established protocols.

There were no shortcuts taken to rush the Covid-19 vaccines into people's arms—at least not in North America. I don't want to seem greedy, but to make a point, I am prepared to stand at the head of the line to receive this immunization.

THE DECAMERON REVISITED

While sitting secure in my home and waiting for the Covid-19 pandemic to pass, it occurred to me that our predecessors have also hunkered down in hope of surviving a plague. I thought of the year 1348, when Mongol traders brought the Black Death to an Italian trading post on the shore of the Black Sea. During the siege, the Mongols catapulted the bodies of their plague-dead into the Italian outpost, which caused the Italians to take to their boats and flee toward home, not knowing that they were already infected. The rat-borne *Yersinia pestis* boarded ship with them, and they took the Black Death across the Black Sea to Europe, where it ignited a plague that, according to different authors, exacted a fifty to seventy percent mortality rate.

The grimness of the situation at that time notwithstanding, there were also attempts at humour and levity, sometimes to the very margins of insanity. At that time, Giovanni Boccaccio, an Italian scholar, wrote *The Decameron*, in which he portrayed a mixed company of ten young men and women who resolved on a measure of "social distancing" by hiding themselves in an out-of-city villa for two weeks while they waited for the plague to abate. They agreed to pass the time by telling stories to each other. Each member took charge of the gathering for a day and each of the ten told a story based on a theme selected by the leader of the day. The tales that Boccaccio produced from the mouths of his characters are known to have been widely told throughout Europe, and his book was essentially a recording of the gallows humour of the day. They are, by turn, inventive, pornographic, and disrespectful of both sacred and secular authority—often all of the foregoing at once. The resulting book might, in modern reprints, be subtitled *100 Dirtiest Plague Jokes*, or something similar, but the content was not much elevated above modern day; think, "Did you hear the story about the Priest, Rabbi, and Politician who dropped into a bar . . . " It should be possible, in the present time, to invent something worth a laugh involving Lady MacBeth in a fast-food restaurant with a dirty washroom, but I'm not holding my breath; or maybe I should. The standards of public discourse in stressful times have not changed much in the past six centuries.

I googled "Covid-19 Jokes" to see if our current crisis is producing anything memorable, but not yet. There is a lame comment from the Duke of Cambridge that produced a giggle, and there are a number of attempts involving hoarded toilet paper, but nothing that bears repeating. Nor is it the moment to invent jokes about our fumbling politicians—not while we are looking to them for leadership to get us to safer ground. We may have to defer searching for humour in this situation until after the hurting has stopped, and we can risk it to laugh at ourselves. In the meantime, just keep on smiling.

THE ENTREPRENEURIAL BRAIN

"How smart it was of sheep to invent shepherds"

– Daniel Dennett

Dennett went on in the talk I heard him give, to explain that in the shepherd, sheep founded an entity to whom they could subcontract all their needs for defense, nutrition, and healthcare; all that at the price of only a slightly reduced list of breeding partners and a slightly greater likelihood of being eaten. In Darwinian terms, it was a successful adaptation, as their numbers greatly increased, and they managed to spread to all continents of the Earth. Their lives were rendered much less stressful, and over time, with less to worry about, the size of domesticated sheep brains decreased compared to that of their wild cousins. By way of a slightly twisted extension of this already-twisted imagery, I find myself wondering whether our brains also take opportunities to offload some cognitive obligations to other systems. Let's look at what may just possibly be occurring.

In a way, we are like sheep in that we have learned to find comfort in our relationships with our pets, especially dogs. It seems to have begun with guide dogs, rescue dogs, and tracking dogs. Lately, comfort dogs have also found a place in the reduction of stresses, whether that be for students before exams, veterans after war, or others who find life difficult.

Does the brain have any means to offload any of its duties to other parts of the body when there isn't a dog equivalent available? Does the violinist's brain fully control her hands when she plays "Flight of the Bumblebee," or might much of the musician's protocol be embedded in the muscle memory of trained limbs? Where is such learning inventoried? I recall a BBC science report concerning learning in earthworms that recovered from being cut in two. In the usual manner of divided worms, both halves survived, grew new heads and tails as needed, and thereby doubled their numbers with previous learning remaining intact. Surprisingly, the tail end that had lost its brain remembered its training as well as the cranial end. It seems that worms store their learning, and perhaps other cognitive functions, in tissues other than their brains.

A woman named Claire Sylvia, who underwent a heart and lung transplant, awoke from the anesthetic knowing the name and having some "memories" of her donor. These were verified by his family. She also began, for the first time, to like the taste of beer, a beverage her donor had enjoyed. Her book, *A Change of Heart,* describes her experience. Along the same vein, other organ recipients are reported to have developed new personality traits post-transplant, somewhat resembling those of their donors.

There is a question staring out at us from the bushes of our alarmed ignorance: does the brain allocate any cognitive functions to non-neurological tissues? We could even go a step further and ask what, if any, functions are sublet to our microbiome. There is evidence that micro-floral changes occur in the gut in relation to mental health status, and we do have other evidence of a close relationship between brain health and gut bacteria. I'm beginning to imagine that the brain, far from being the totalitarian custodian of our personhood, is something more like a moderator of a cognitive repertoire that is more broadly distributed throughout our bodies. There is much here to be wondered at.

THE UNIVERSE'S FIRST BABY PICTURE

Dial back with me to the days of analogue television. I'm sure it has happened to you that you sat down after dinner to watch a movie with your family and fell asleep before the end. When the movie was over, the family tiptoed out of the room, and you awoke after the national anthem and the test pattern to see electronic snow on the screen. Most of the so-called snow was due to the noise inherent in the electronics of the time, but about one to two percent was coming at you from a universe that was a mere 380,000 years old when the signal was transmitted. This was the Cosmic Microwave Background (CMB).

In the 1960s, a couple of telephone company engineers, Arno Penzias and Robert Wilson, were trying to understand radio static. It was the era of the moonshots, and clarity of communications was critical to mission success. Bell granted them access to a radio telescope at their lab in Camden, New Jersey, which the investigators used to search the sky for possible sources of radio noise. This was all to no avail because there was a faint signal coming at them from every direction. It was suggested they speak to Bob Dickie, an astronomer at nearby Yale University who was hoping to test a theory that there should be a signal still detectable from the cooling of the early universe.

Dickie took the telephone call while he was in a teaching session with graduate students. He listened to Penzias and Wilson, offered an explanation, and, as he put the phone down, said to the students, "Boys, we've been scooped." Penzias and Wilson, who hadn't really understood what they were doing, nevertheless received a Nobel Prize for their discovery.

The all-sky image of the CMB, taken from the Planck satellite is viewable from the gallery of the European Space Agency. It is an image that is rich in information. Among other things, it tells us that the universe had a birthday. It hails from the first instant in time after gases in the universe had cooled enough to transition space from an opaque plasma to a transparent vacuum. The colours in the image show that the temperature is remarkably uniform, since the hottest (red) to the coldest (blue) spans a range of one hundred thousandth of one Kelvin.

A second revelation hidden in this image is that there are 180 degrees in a triangle—not just any old triangle but one drawn on the entire sky. Before this revelation, there were several competing geometries for the shape of the universe, some of which would have had you seeing the back of your head on the horizon. This insight allows us to use a flat geometry to study the universe.

A third observation has yet to be established with data, but it follows from the polarization of the CMB and predicts that we will learn from the gravitational distortions induced in the first instant of the Big Bang, when inflation enlarged the new universe at a rate that exceeded the speed of light. The technology with which to achieve those observations awaits the construction of new technology, such as LISA the space-based gravitational wave detector.

You could have worked this out for yourself except that you fell asleep during the movie.

THE GRADUATE STUDENT

He's back. I'm surmising that he is still a graduate student who has been sent to collect data for his mentor again this spring. I'm speaking of the male cardinal who, for three summers now, has spent mornings attacking his silhouette in my windows. In my imagination, he is studying the impenetrable mysteries of invisible matter as dense and unknowable to his bird brain as dark matter is to mine. He made his first trial flight of the season at 06:25 this morning, thumped himself against my window and distracted me from my reading. If the past serves to predict, he will continue with this daily data collection until midsummer, and I hope for him that he will get to writing up his thesis before he suffers a major injury.

I am reading Daniel Dennett's book, *From Bacteria to Bach and Back*. It is about the evolution of mind from an unconscious beginning. I have been digesting his insights around the development of competence without comprehension—that is, creatures learning to do things without knowing what they are doing. His examples include the construction of elaborate nests by insects and birds who seem to lack the cognitive machinery that would allow them to vary their efforts in the manner of cognitively aware architects. If Dennett is onto something, then my "graduate student" also has no idea what he is doing. Will he ever learn?

On reflection, I think that I have also experienced competence without comprehension inside my own head. Many years ago, when long distance telephone calls were very expensive, my hospital installed a telephone password system to block frivolous use of the switchboard. At the time, I was internationally active in my specialty and in need of daily worldwide telephone access. The hospital gave me a seven-digit password with which to precede each such call. For a couple of years, I routinely held the handset in my left hand and dialed with my right. That is, I did this until one day when I came to work with a very painful right tennis elbow that precluded reaching across the desk to the phone. When I attempted to dial with my left hand, I discovered that I no longer knew the password; it seemed as though the code had gradually moved out of my conscious memory into my right hand and then become cognitively unavailable. I had become competent but uncomprehending!

It seems then that my cardinal and I may not be so different in this respect, as neither one of us fully knows what we are doing. Perhaps we should conduct a workshop in which we will each teach the other.

THE GREAT DISCOVERY
THAT NEVER WAS

I have just read Guido Majno's 1975 book, *The Healing Hand*. Majno was a mid-twentieth-century pathologist at Harvard who, in mid-career, seems to have left the practice of pathology to learn to read the records of medical practice in the ancient Greek, Egyptian, Syriac, Arabic, and Mesopotamian languages. It's not so clear whether he also managed to read some of the ancient Chinese prescriptions. The focus of the book is on the management of wounds, both in the acute and actively bleeding phase through to the infected and purulent. On those occasions when he identified potentially falsifiable ideas and practices from history, he took them to his laboratory for further testing.

The Egyptians of about 3,500 years ago applied mixtures of honey and rendered animal fats to prevent and treat infection in wounds. In Majno's lab, both honey and fats proved to have antibacterial properties. About twenty-five years ago, my colleague, Dr. Robert Lannigan, conducted a similar experiment on the effect of honey on bacterial cultures. We found that multidrug-resistant Staphylococci (MRSA) were inhibited by honey freshly recovered from my apiary, but not by honey that had been in storage for a few months. Pasteurized commercial honey also had no inhibitory effect. However, a pot of honey, purchased on a visit to the Hadhramaut region of Yemen, retained antibacterial properties, even after several years. The Egyptian physicians were onto something, but they never took it to the point of experimentation. I learned from conversations with African physicians that to this day, some African surgeons still apply honey prophylactically to fresh surgical wounds.

For my part, having spent a part of every summer weekend in my apiary for thirty years, I can attest to the fact that bees are very clean. In a springtime hive, the combs at the edges beyond the winter ball of bees would initially be green with the overwintered microbial growth. As they began to prepare for summer, the bees would clean up the combs until they were pristine and white. To do this, young bees were engaged to carry all the debris out of the hive, and the salivary excretions left on the combs killed what was left. What the bees bring back to the hive is not honey—only a dilute, slightly sugary solution of nectar that has to be progressively evaporated to achieve the final product, whose osmotic pressure is also antibacterial. To achieve this dehydration, the young bees anchor themselves in position, all rears pointing to the exit, and beat their wings in unison to move air through the hive day and night. They augment the evaporation of water by moving the nectar from cell to cell in the comb and maximizing the exposed surface area. In this process, they also deposit more antibacterial enzymes in the honey, since in their economy, it had to remain edible, at least to the end of next winter. Archeologists have recovered honeycombs from Egyptian tombs from two thousand years ago, and they report that it still smells like honey—another testament to its resistance to bacterial decomposition.

Unfortunately, in that early era of food production preceding publications and conferences, the value of honey in wound treatment wasn't told to the outside world, and neither the Greeks nor the Romans took it up. Hippocrates in Athens never mentioned a role for honey in the treatment of infection. The reputation of an aspiring physician in those times was often determined by the novelty of his therapies and the aggressiveness of his marketing. Many wound treatments had toxic and revolting ingredients that were typically compounded in secret and included animal excrements and the like. Small wonder that infection came to be the expected outcome of wound-healing! An unimaginably large number of people with initially clean wounds must have succumbed with lethal infections as a result of their treatments.

There are still places around the world where wounds are treated in unscientific ways. The first time I visited Taiz, I was taken on a tour of the Swedish-Yemeni Children's Hospital. There were people there who were focused on the management of malnutrition in children, and they were doing a good job in their corner of baby care. Then, I was profoundly shocked when they took me into their neonatal tetanus unit, where there were some five or six newborns with tetanus. The families were all sitting on the floor, being very quiet, and the lights were turned down to avoid stimulating the babies into exhausting tetanic spasms. I was there to teach a course to senior doctors on hospital management, and the tetanus ward became a subject of discussion. In discussion with other Canadian doctors who had been to Yemen before me, I learned that untrained country midwives were accustomed to applying a dab of donkey dung to the freshly cut umbilical cord to stop bleeding and that was a potent source of the infection.

I revisited that hospital a year later and was so pleased to see the changes they had made. The department chief was a strong leader who had initiated development of written and evidence-based tetanus therapy guidelines, including muscle relaxants that reduced the tetanic spasms. This innovation enabled medical teams to provide a consistent level of care despite shift changes. The ward was better lit, so nurses could see to provide care. Parallel to that experience, there were some encouraging encounters with schools of midwifery, which were trying to bring modern concepts to the profession throughout the country.

One more reflection about Majno's intellectually enriching book: with all the ado surrounding the mixing of ingredients in the various potions, lotions, and toxins that were proclaimed as medicines, how was it that no one seems ever to have questioned the quality of the water that was used to make them up?

There was a mythology propagated in the Roman times that water was toxic, and people took to drinking wine and beer to avoid cramps and diarrhea.

How was it that these various intelligent physicians and others never asked questions about water quality? They came so close when they recognized the preferability, at least in the short term, of alcoholic beverages over water. They were only one question away from asking why water causes diarrhea, or from quality assuring the role of water in their medicines. In the absence of that critical question uncounted millions of the Old World's inhabitants perished so miserably of cramps, dehydration and infection.

THE LANGUAGES OF BABEL

Linguists have long been intrigued by the relationships among language groups such as those of the Indo-European family. Across several hundred languages, similar meanings often appear as similar sounding words; take for example, *water*, *vasser*, *vota*. From these modern similarities, we can make educated guesses as to what may have been the original word. One of the most studied sets of terms relates to wheels and axles with animal-powered transport, which first appeared about six thousand years ago in the approximate region of modern eastern Turkey. With the increased mobility afforded by wagons, languages also moved, and words denoting new things came to be shared across cultures. Not that there is an Indo-European root for "automatic transmission," but more likely for wagons, harnesses, and the surrounding technology. Over time, these and many other words spread, and in response to shifting linguistic needs, they mutated in subtleties of meaning and their sounds. Nevertheless, it remains possible today to trace back some of our modern words to their places and times of origin. More examples could be developed with the terminology of beekeeping and cloth weaving.

Something similar has also happened over the past few decades in relation to our understanding of DNA sequences. If I and a cousin find that we share a rare genetic variant, then we can be certain that it came from a grandparent who lived in a certain place. By extrapolation, we can estimate how many of our descendants also share the gene, and by examining dispersion of the populations, we can predict where else this gene might turn up. Further, by extrapolating backwards and checking against the sequences, we can now read ancient DNA, we can posit the genetic makeup of our earliest pre-human, ancestors. Like our Yemeni limo drivers' use of language, all humans have borrowed useful genes, such as parts of our cellular immune system, from our Neanderthal cousins of 50,000 years ago. We appreciate now that life is its own language of which DNA is the archive as well as the operating manual. The language of DNA has an alphabet consisting of sixty-four letters with rules for marking the beginnings and endings of genes, as well as for editing and declaring exceptions, correcting typos and much else. In a real sense, we know the "DNA words" of our first ancestors with more certainty than linguists can know the original word for a wheel.

One example of how mutations alter the wording of the genetic language of humans is in the appearance of tolerance to milk drinking into adult life. Before our ancestors took up lifestyles that included herding of cattle and consumption of milk products, all humans were lactose intolerant in adult life. In a plains-dwelling society, with periodic food shortages, milk became useful, at first to support the nutrition of children and only a few adults. The discovery of cheese became a way to store milk fats and proteins for a longer term. Over a relatively short time the advantage of lactose tolerance selected for survivors although a minority of humans remain lactose intolerant into modern times. In short, a change in lifestyle promoted a change in diet and brought about an amendment to human genetic expression.

I find it remarkable that life and language should imitate each other so closely. There must be more than a coincidence at work here.

THE MARVEL OF MINERALS

I was going by a pharmacy cabinet that displays mineral supplements when I was overtaken by a rogue thought. These were not merely nutritional supplements that I was seeing, but a museum display of fossils that life internalized into what became plant and animal bodies when life originated four billion years ago, when metabolism was still no more than a tentative experiment. Those refined salts, now on sale for gulping with your breakfast, represent the rough under-water crystal surfaces on which the first layers of pre-biological chemistry were learning how electrons could move to make the effort worthwhile. Lest you wonder what I am getting at, look up Robert Hazen's course on geology in The Great Courses on YouTube. We haven't had nearly the close contacts of biology with geology that the subject of life's origins demands. Those pristine crystal surfaces now in the bottles (of calcium, iron, magnesium, copper, zinc, and others), served in crystals as the anvils on which the first organic molecules were shaped to variously make or break molecular bonds and flattened the thermodynamic barriers to biochemical reactions, thereby channeling the support that evolving life needed. If you understand electron tunneling in the circuits of your cell phone, you have discovered the essence of what enzymes are doing for you in your cells.

In other words, those mineral crystals were of the correct shape to be the first enzyme prototypes of what today directs and energizes the fabric of life. Modern enzymes still contain those ions and remain capable of nudging electrons along desired paths to do work as needed by cells. All the cells in our body—all approximately one hundred trillion of them—can marshal thousands of unique enzymes each, and many of them still have a mineral atom in a location that is critical for their function.

I wonder whether the other customers are also impressed by the marvel of minerals.

THE TWO-BRAIN DILEMMA

You might not know that you actually have two brains. There is the old brain (paleocortex), inherited from our pre-dinosaur ancestors a hundred million years ago. That brain more or less resides in the brainstem, cerebellum, basal ganglia, and spinal cord, and it manages much of the day-to-day administration of our lives—heart and respiratory rates, appetites, temperature regulation, alarm systems, and the like—without our having to think about such matters on a minute-to-minute basis. The paleocortex is the source of much delight and great desperation. On the one hand, it drives appetites and passions of every kind; on the other, it is also the source of impulses that drive fear, hatred, and aggression. The paleocortex does not seem to have any built-in mechanisms for the consideration of consequences, guilt, or regret. However, on those occasions when a split-second decision is called for, it is the paleocortex that steps up. Depending on the actions triggered, it is the maker of both heroes and villains. You may look silly if you dive for cover on account of a rustle in the grass behind you, before realizing it was only a passing breeze, but it might just possibly have been a lion preparing to pounce on its lunch! False positive reactions are tolerable when risks are high.

There is also a new brain (neocortex), much of which developed over the past five to six million years; it resides in the cerebral hemispheres, and it allows us to anticipate consequences, think in abstract terms, communicate with others of our kind, reason, justify motives, accept responsibility, acknowledge guilt, and entertain ethical dilemmas. In short, while the paleocortex drives lust, the neocortex has the means to adjudicate the conditions for love, among other things.

The problem is that the two brains are not completely integrated, though they both demand to be exercised in integrated day to day life. Sports play an important role in the maintenance of the paleocortex, but the neocortex is the referee, and the post-game neocortex assures that further combative behaviour is reined in over a beer. Nevertheless, the old brain is easily subject to manipulation when, for example, fear becomes an element of any situation. We are primordially afraid of the stranger who might just be intent on dispossessing us of our cave, our followers, or our goods, and we justify resistance, even violence, on grounds that we were afraid. That is also a reason why political debates so often resort to the invocation of fear in the place of reasoned debate. The new brain, on the other hand, continually works to mitigate fear and uncertainty through systems of philosophy, laws, procedures, courtesies, and protocols.

We humans are seriously messed up. The modern caveman forced out into the light of artificial intelligence and nuclear science looks a lot like a toddler running with scissors. But how can an accident be avoided? I certainly have enjoyed many of life's most passionate moments and would never give them up. However, the current round of political preachings that are based on the exploitation of fear and the exclusion of strangers is exceedingly dangerous. Perhaps it helps to recognize that the problem not only surrounds us, but also lies within us, and each of us must seek to make peace between our two brains.

THE SHAPE WE ARE IN

There is a phenomenon in biology known as enantiomorphism. It might be a good word to practice today. What it means is that carbon atoms have four valence electrons; if these positions in the atom are occupied by four different end groups, then the resulting molecule is asymmetrical in the same way as left and right hands are asymmetrical. This is important for the configurations of proteins and carbohydrates because they need to fit into active sites of the correct handedness on enzymes, rather the way that a right hand can only fit into a right glove. We can measure the degree and direction of enantiomorphism by the rotation of a beam of polarized light as it passes through a pure solution of the compound in question. The scientific shorthand for these enantiomers is l- (left) and d- (right).

Whenever we synthesize a carbon-based compound in the laboratory, we produce a mixture of equal amounts of both l- and d- compounds, and there is no net polarization of the product. However, all compounds made by a living cell contain only l-amino acids and only d-sugars. The reverse-handedness has been sought, but not found, in living matter on Earth. We are now satisfied that such mirror-life does not exist here, though the reasons for this remain obscure. Hang on to this mystery for a bit while I pull another thread through the fabric of my thought.

One can look in many places for the molecules that form the basic building blocks of life, and they seem to be abundant wherever sought for. Infrared spectroscopy performed by space telescopes identifies sugars, fatty acids, and amino acids in the nebulae of dead stars in outer space. These same compounds can also be extracted from meteorites that came from space and crash landed on Antarctic ice. It might be argued that meteorites crashing onto the Earth could be contaminated by contact with our planet's surface on their way to the laboratory and, to this end, the Japanese rocket, named Hayabusa, collected clean specimens from space in a snatch and grab exercise and successfully returned samples to Earth for analysis. These samples also contained the same molecules of life, with the additional information that they were enriched with respect to the l-amino acids. The space probe, named Rosetta, landed on the tip of a comet, and the materials it recovered yielded similar results, confirming the seeming excess of l-amino acids in space. I don't want to think that this is just a coincidence.

I don't know anything about the geology of space rocks. I am tempted to speculate that, as is the case on Earth, there are also enantiomeric rock crystals in asteroids and comets, and that these crystals may have been the "anvils" on which those possibly original organic molecules were formed with the energy possibly provided by ultra-violet light. It is possible that the handedness of living nature may have been laid down in outer space before Earth was formed. Thus, one can predict that when life is found on other planets, it will also be constructed from l-amino acids and d-sugars. Carl Sagan liked to say that we are made of stardust; there may be more to that than even he knew.

A HERO OF THE PIONEERS

There must have been a sense of expectant exhilaration in the early pioneer communities of the prairies, as newcomers learned to plow their sod and turn it into productive fields. Sure, the beginning was tough, but the prospects were so great. Many of the settlers, such as the Ukrainians, Mennonites, and others were simply relieved to have escaped blood-sucking tyrannies and were, in the first instance, ecstatic merely to breathe the air of freedom. Others had fled personal demons, and their memorable characters added colour to the emerging social fabric. There was a strong contingent of settlers who were drawn by the prospect of becoming landowners. Imagine how prosperous everyone will be when the wheat fields stretch all the way to the horizon! Markets seemed assured because the Government of Canada had shown the insight to build a basic economic matrix in the form of transportation (the railway) and law (the RCMP, née RNWMP) before opening the region for settlement. So, what could possibly go wrong? Over time, whatever could go wrong, did.

Given the environmental unawareness of the day, no one anticipated the impact of converting a large part of the continent from a heterogeneous equilibrium of prairies and woodland into a monoculture of wheat—three whole provinces of it. The new environment proved an ideal support for plagues of grasshoppers that could destroy entire crops in a few days. Rainfall was erratic, and crops failed on occasion. Dry times were complicated by dust storms and erosion. I can still recall the taste of dust on my teeth even though my mother kept me indoors during those days. Despite closed doors, the house leaked enough to permit dust penetration on a large scale; every surface became coated. There is hardly a more dismal sight in my memory than the kitchen window view of the endless Russian thistles cartwheeling past the kitchen window and through the yard, catching up in the fences and the dust blizzarding into drifts.

And when everything seemed to be going just right, when the Sun was shining and hearts were full of hope, one extra rainfall before harvest might provide the moist milieu needed for plagues of rust fungi to thrive and overnight rendering bumper crops of wheat unsaleable and fit only to be burned.

The growth of wheat for the international market was to be the backbone of the emerging economy. The strain brought from the east was known as Marquis. Marquis wheat was excellent for breadmaking, but it was also exquisitely sensitive to rust fungi. When it was planted in patchwork quilts of hundreds of acres stretching across the horizon, it proved an irresistible nutrient for the fungus as the spores were blown across the plains by the wind.

By the nineteen thirties, after a couple of bad years, people began to question if they really had a future on the prairies so long as this plague persisted. A solution was urgently needed. Dr. Saunders, a young professor in the Department of Agriculture at the University of Saskatchewan, undertook to study the problem. The story we were taught in school was that he began by simply walking through the fields to survey the dimensions of the problem. At first, it all must have looked the same: once-bountiful wheat crops turned into a worthless matt of grey garbage, taking with it all the work and treasure so invested. But eventually, he began to see a pattern developing. Here and there, at long intervals, were single stalks of wheat that had seemingly shrugged off the blight and remained perfectly healthy. These individual stalks he took back to the

University's greenhouse, where he determined that their resistance bred true through subsequent generations. By the mid-thirties, we had the Saunders strain of wheat that was resistant to rust and smut fungi. Thus, the Prairies were saved for wheat growing. Professor Saunders's work was held out to us in our schools as heroic. I recall my father talking about it with neighboring farmers; it seemed that we could now relax a bit, having beaten back an enemy who had threatened our very survival.

We didn't see Dr. Saunders' achievement then through the eyes of evolutionary biologists. We didn't think past our own immediate survival, and this left no room for ethical concern for the consequences of genetic manipulation. Modern geneticists would have solved the problem at the molecular level in less time by manipulating the appropriate genes—a process that would be criticized by our current generation of "back to nature" purists—but the outcome would have been the same.

I remember the relief in my father's voice when he and the neighbours spoke about how close they had all come to disaster over the recurrent rust plagues.

THE SUPPLIANT MAIDENS

A few years ago, I attended a performance of a Greek drama; I don't recall which one. It had an interesting dramatic structure, with a storyline built up by the lamentations of a solo performer who was backed up by a more contemplative chorus that interpreted the events and offered social commentary to the audience as the play progressed. That experience caused me to take more interest in this literary genre, but life was in charge of my schedule, and I didn't get around to reading another Greek play for a long time thereafter.

Recently, I came upon Aeschylus's work, *The Suppliant Maidens,* which was written about 2,600 years ago. I remembered how much fun this could be. The Greek sense of plot and drama, even so long ago, was highly developed, and it remains obvious to modern readers. Briefly, this play tells the story of a family living in Egypt. The family has twelve sisters, each more lovely than the last, who are being aggressively pursued by unwelcome Egyptian suitors who become threatening and aggressive when the women reject their advances. Fearing for their futures, the sisters board a boat with their father and make their escape to a Greek island to avoid lives of marital enslavement.

No sooner have they landed that they discover that they have been followed across the Mediterranean by their bullying suitors, and that they might now be trapped on the island with them. Dramatically, they plead with the local authorities to protect them.

There follow scenes that deal with the rights of refugees across international boundaries, the ethical limits of diplomacy to force return of unwilling refugees despite a threat of war between Greece and Egypt, the ethical duties of host societies to behave with courtesy toward desperate humans, and women's rights. Were it not for the stilted medieval language of my version of this play, it might have been written in our own time, as it appeals to our own ethical instincts about the duty of care and the rule of law in society toward our own distressed population groups.

As often as I think about it, I am again amazed by the mere fact that we have such ancient treasures still with us. Greek literature was, of course, written in Greek. In the subsequent centuries the value of the books was played down because they were not "Christian." The lamb skin palimpsests on which they had been written were often recycled by monks in search of materials on which to write new Christian works, such as prayer books. (It may have been here that "handbooks" originated because the monks typically would fold the large, erased pages of old books to make two of a smaller volume.)

In the European Dark Ages, Greek literature of all kinds was devalued. Then in the age of the Caliphs in the Islamic Middle East (eighth century), book merchants from centres of Islamic scholarship who were mandated by the likes of the Caliph of Baghdad, purchased as many of the European books as people would part with, and took them to Baghdad, where they were translated into Arabic for local study. Centuries later, as the West began to reawaken, many of these works were, out of courtesy, retranslated back into the original Greek and Latin and returned to the new generations of Western scholars. The play I read may have been through this unlikely labyrinth, and it is no small wonder that it has survived at all. Not only the plays, but also many Greek philosophical treatises have, through time, informed and influenced Christians and Muslims as well as Greeks. Let no one say that these words were wasted.

I admit that not all the Greek plays are as accessible to the modern reader as *The Suppliant Maidens,* but if you want to dip a toe into these waters, this is a good place to start. There is so much more to this than a mere story.

THEY CALLED IT PERCY

Recently, NASA successfully landed its newest rover, Perseverance, (aka Percy), on Mars. The landing alone was impressive; four months into a six-month flight, a single mid-course correction was made, which brought the lander down within feet of its target on the old martian river delta without further human intervention. The next goal is to explore the area, find the preferred places to drill soil cores that are to be cached for pickup by a courier mission in 2031, and returned to Earth for examination for evidence of life. I hope I have ten more years so I may also learn of that result.

Until about seventy years ago, hypotheses around the origin of life were entirely speculative, with little prospect that any would ever become testable. Then, in the 1950s, Stanley Miller made an experimental atmospheric model to test possibilities of conditions on the early Earth, and he found that atmospheres containing only water vapour, methane, carbon dioxide, sulfur oxides and energy from simulated thunderstorms would, in a few days, create a mixture of tarry materials that contained amino acids, sugars, fatty acids, alcohols, and other compounds of relevance to life. Such lab experiments, together with spectroscopic observations of nebulae in the night sky, taught us that the basic Lego building blocks of life's chemicals were readily available on Earth before there was life; they were also scattered throughout the universe, where there may or may not also be life or an analogue of it.

270

But where should we look for evidence of the mechanisms by which life's chemicals could be brought together to make organisms? We have only the example of Earth to learn from, and the time has come to study it closely because Earth may provide us with the most likely example of early conditions on other planets. The origin of life has become the most urgent question in all of science, and there is now a distinct possibility that answers will be forthcoming.

From what we know so far about early events on Earth, and what we have learned recently about the formation of planetary systems, it is probable that similar conditions for life also happened on Mars, and even on exoplanets. Therefore, it is unlikely that life elsewhere, if it exists, will be very different from life on Earth. Thus, the best way to prepare for the science of the future is to study ourselves in the here and now. That is why Percy was not outfitted with an elephant gun, and a pith helmet to search for life; instead, he was given a core drill to extract samples of the inside of Mars. The results, however they come out, will have profound implications for our understanding of ourselves and our likely origin on Earth.

The Kepler satellite studies, and other probes still collecting data, assure us that planet formation is a routine part of the development of every star. But do any planets nurse the conditions that we might think of as providing opportunities for life—i.e., water, temperature range, atmospheric conditions? For most cases, we do not yet have telescopes with the resolving power required to answer these questions, but some now under construction will be able to do that within this decade—e.g., the Thirty Meter Telescope. To this end, the field of astrobiology is becoming recognized as a proper field of study at several universities, and if some of our grandchildren should choose to become astrobiologists, don't talk them out of it.

'TIL WE ALL HAVE FACES

As I write this, it's been more than five weeks since I saw a human face other than that of my son, who brings me groceries. Last week, I had an appointment at University Hospital, where all my caregivers were masked as part of the protocol of the Covid-19 pandemic.

The masks took my mind back to the times I worked in the Middle East where women, with few exceptions, were required to be masked. Over the nearly twelve years I visited there; I delivered about fifty lectures to Middle Eastern hospital medical staffs. Routinely, the veiled women doctors would arrive early and seat themselves at the back of the room in a quiet huddle. Later, the male staff would come in, sometimes accompanied by one or two unveiled women who sat with the men. In all those lectures, I never fielded a question or received a comment from a veiled woman, whereas the few that were unveiled provided regular and powerful critiques of whatever I was teaching. It was similar in clinics with the medical students; veiled female students made little comment in contrast to the chatty ones with open faces. I observed that taking away someone's face is a very effective way to shut them up and take away their personal power.

At the King Faisal Hospital in Riyadh, I once provided some merriment for the women in the department I worked in because they were all of similar build, and I wasn't able to distinguish among them from behind their veils. Eventually they revealed that they worked diligently to give subtle clues to their individual identities within a narrow range of options. One of them wore her mask with the upper edge at the level of her lower eyelids, whereas her colleague dared to display three millimetres of skin.

One of my colleagues in Yemen was a Canadian gynecologist who was sought-after for his expertise with female reproductive system cancers. My colleague told me about his practices and the ways he sought to engage directly with his patients. The practice of Arabic physicians was to provide a diagnosis to the husband or other accompanying male without speaking a word directly to the woman. My colleague insisted on speaking directly, and for this he would ask her to first remove her veil. When she demurred on grounds that it was un-Islamic and immodest to do that, he would explain that he needed to see her face to have the conversation she deserved, and that he didn't buy the argument about modesty because she had not been overtly embarrassed to show him her bare bottom just minutes earlier. In fact, he was trying to make a point to the Yemeni doctors in attendance that female patients deserved the courtesy of direct discussions that a more veterinary style discussion with the husband did not address.

After working in this environment for a few weeks on multiple occasions, I noticed a profound relaxation in my state of mind when we stopped in Frankfurt on the way home. It had everything to do with the appreciation of mixed crowds, including women with faces. I just felt the relief that came with the appreciation that I was back in a society where people were valued as self-expressing individuals regardless of gender.

Through the recent pandemic, there has been a vocal subset of society who objected to wearing masks. I think I understand their resistance to the idea that when masked, they are not so much in control of their destiny. The mask says to these folk they are not so much in control of their bodies as they want to imagine, and I might agree. They are not appreciating the reality of the virus enough to give up, even for a short time, on this vital item of their identity. I admit that faces are important.

There is a place in the human brain where faces are uniquely recognized, and the memory of them is stored in a way that other body parts are not. Prosopagnosia is a condition in which faces are not remembered. It may be a birth defect, as may have been the case with Oliver Sacks, the late neurologist and author. Apparently, he could work all day with colleagues and then fail to recognize them in the parking lot on the way home. There is also a rare form of stroke that produces this effect. In these cases, the masks exist in the mind of the observer rather than on our faces.

When we can, we should rejoice in the power we share and that becomes meaningful when we show each other our faces.

CODA

Salman Rushdie's novel, "Victory City" concludes with the demise of his heroine/demigoddess, Pampa Kampana, and the erasure of her city from history. Having come to the end of her story with its moments of glory and sieges of desperation, and now with nothing to look forward to, she makes her own most courageous statement; "Nothing survives. But nothing is meaningless either."

All the lines of our experience of the universe are threads, both beautiful and often simultaneously also painful. that together make up the glorious fabric on which our stories are told. What does it all mean? Does it mean anything? Does materialistic science rob us of the possibility of meaning?

I think that the universe unthinkingly has left open an opportunity for each of us to write the sections that deal with our personal meaning of life. No greater opportunity could be offered than to write our own chapters and drop them into the universe's Table of Contents. Our different ways of thinking properly add enrichment and contrast to the whole story.

I've shown you evidence of the beautiful universe through glimpses into astronomy, physics, chemistry and biology, all supported by evidence as I came upon it. I've lived long enough to have explored other hypotheses concerning our origins, but the materialistic version alone has proved enduringly satisfying. If there is more, let it be so but, it's all I can vouch for. That is enough to fill me with joy and give me hope. Pampa Kampana would, I think, agree as well.

The writing forced me to rethink and restate some of my own most basic premises and I hope you also enjoyed the book.

Al Driedger